Mechanics of Failure Mechanisms in Structures

SOLID MECHANICS AND ITS APPLICATIONS
Volume 187

Aims and Scope of the Series

The fundamental questions arising in mechanics are: *Why?*, *How?*, and *How much?* The aim of this series is to provide lucid accounts written by authoritative researchers giving vision and insight in answering these questions on the subject of mechanics as it relates to solids.

The scope of the series covers the entire spectrum of solid mechanics. Thus it includes the foundation of mechanics; variational formulations; computational mechanics; statics, kinematics and dynamics of rigid and elastic bodies: vibrations of solids and structures; dynamical systems and chaos; the theories of elasticity, plasticity and viscoelasticity; composite materials; rods, beams, shells and membranes; structural control and stability; soils, rocks and geomechanics; fracture; tribology; experimental mechanics; biomechanics and machine design.

The median level of presentation is the first year graduate student. Some texts are monographs defining the current state of the field; others are accessible to final year undergraduates; but essentially the emphasis is on readability and clarity.

For further volumes:
http://www.springer.com/series/6557

R.L. Carlson • G.A. Kardomateas
J.I. Craig

Mechanics of Failure Mechanisms in Structures

Springer

R.L. Carlson
Georgia Institute of Technology
School of Aerospace Engineering
Ferst Dr. 270
Atlanta, GA
USA

G.A. Kardomateas
Georgia Institute of Technology
School of Aerospace Engineering
Ferst Dr. 270
Atlanta, GA
USA

J.I. Craig
Georgia Institute of Technology
School of Aerospace Engineering
Ferst Dr. 270
Atlanta, GA
USA

ISSN 0925-0042
ISBN 978-94-007-9819-9 ISBN 978-94-007-4252-9 (eBook)
DOI 10.1007/978-94-007-4252-9
Springer Dordrecht Heidelberg New York London

Printed on acid-free paper

Springer is part of Springer Science+Business Media (www.springer.com)

Preface

As manufacturing and transportation technologies have evolved and become more complex, the ranges of the operating conditions in structures, machines and vehicles have expanded. The consequences of failure in these systems can range from being minor to catastrophic events. The importance of the development of a basic understanding of the interactions of complex failure mechanisms is clear. The goals of the experimental investigations of failure mechanisms are also clear. They are to provide foundations for the development of analytical procedures that provide for the development of sound design methods, and effective monitoring procedures. Mechanisms that can lead to failure include fracture, fatigue, creep, collapse due to buckling, and dynamic excitation.

The emphasis is on failure mechanisms in metallic, ceramic, polymeric, and composite solids. Chapters are devoted to failure mechanisms that are characteristic of each of the materials. Failures in Chap. 1 of metals include time independent and time dependent tensile and compressive loading for both static and dynamic cases. Both theoretical and experimental results are presented. Chapter 2 includes a consideration of the basic elements of fracture mechanics, the effects of toughness gradients, and the effects creep induced brittleness.

The statistical, bimodal character of multi-site fatigue cracking in the small crack regime is discussed in Chap. 3. The use of the range of the stress intensity factor in computing the rate of fatigue crack growth rate in the long crack regime, the effects of variable amplitude loading including tensile and compressive overloading, and environment effects are also discussed.

Failure mechanisms in ceramic composites, polymeric composites and metallic composites are discussed in Chaps. 4, 5 and 6. Failures in biomaterials are discussed in Chap. 7. Failure mechanisms in systems with multiple structural elements are discussed in Chap. 8.

Professor Carlson would like to acknowledge the support and encouragement of Dr. C.J. Beevers of the University of Birmingham in England with whom he conducted research on fatigue. He would also like to acknowledge the benefit of his collaborative research with Prof. N.J. Hoff of Stanford University on the topic of structural stability. The second author, Prof. George A. Kardomateas gratefully acknowledges the benefits of being introduced to research in mechanics by Prof. F.A. McClintock of the Massachusetts Institute of Technology, and he would also like to acknowledge the long-term support and encouragement of Dr. Yapa D.S. Rajapakse of the Office of Naval research. In addition, he would like to acknowledge the benefit of stimulating scientific interactions with Profs. George Simitses, Victor Birman, Leif Carlsson, Izhak Sheinman and Shuki Frostig.

We acknowledge the contributions of Dr. Marcus Cappelli who conducted the experiments and analyzed data for our investigation on the growth of multiple-site fatigue cracks. We also want to cite the contributions of Tina Babanari who participated in our experimental fatigue crack growth investigations.

Finally, we would like to extend our thanks to Ms Nathalie Jacobs of Springer (Dordrecht, the Netherlands) for her help in making this work possible.

We hope that this book will enable the specialists who encounter questions regarding the failure mechanisms of structural elements to have the basic material and overview on this topic at their fingertips and to utilize it in their research or engineering practice.

Atlanta, Georgia

R.L. Carlson
G.A. Kardomateas
J.I. Craig

Contents

1 Stability in Metallic Elements 1
1.1 Time Independent Tensile Loading 1
1.2 Tensile Creep Rupture 4
1.3 Time Independent Compressive Loading 4
1.4 Time Dependent Compressive Loading 5
1.5 Static Buckling Experiments 6
1.6 Three Dimensional Problems 8
1.7 Dynamic Stability 9
1.7.1 A Plate Problem 9
1.7.2 A Tensioned Bar Under Eccentric Loading 10
References 10

2 Fracture Mechanics 13
2.1 Introduction 13
2.2 Fracture in Metals with Toughness Gradients 15
2.3 Metallurgical Embrittlement Effects 15
References 17

3 Fatigue in Metals 19
3.1 Introduction 19
3.2 The Stress-Life Strategy 21
3.2.1 The Stress-Life Diagram (S-N Curve) 22
3.2.2 Mean Stress Effects – The Goodman Relationship 23
3.2.3 Variable Amplitude Loading – The Palmgren-Miner Rule 25
3.2.4 Fatigue Design Approaches: Safe-Life, Fail-Safe and Damage Tolerance 25
3.3 Fatigue Crack Growth 28
3.4 Crack Initiation Mechanisms 30
3.5 Multi-site Fatigue Cracking 30
3.5.1 Multiple Site Cracking Test Results 31
3.5.2 Transition from Small to Large Fatigue Cracks 33

3.6 Variable Amplitude Loading 34
3.6.1 Tensile Overloads 35
3.6.2 Compressive Overloads 35
3.6.3 Load Spectra Analyses 35
3.7 Temperature 36
3.8 Corrosion 37
3.9 Creep and Fatigue Interaction 37
References 38

4 Ceramic Materials 41
4.1 Introduction 41
4.2 Ceramic Matrix Composites 42
4.2.1 Fabrication 42
4.2.2 Fracture and Fatigue Strengths 42
References 43

5 Polymeric Materials 45
5.1 Fatigue Failure 45
5.1.1 Matrix Cracking 45
5.1.2 Delamination 47
5.1.3 Fiber Breakage and Fiber-Matrix Interfacial Debonding 49
5.1.4 S-N Relations 49
5.2 Environmental Effects on Composites 50
References 54

6 Metallic Matrix Composites 55
6.1 Introduction 55
6.2 Experimental and Analytical Results 56
References 58

7 Biomaterials 59
7.1 Introduction 59
7.2 Prosthetic Heart Valves 59
7.3 Prosthetic Hip Joints 60
References 61

8 Failure in Structural Systems 63
8.1 Truss Failure 63
8.1.1 An Isostatic Truss Example 65
8.1.2 A Hyperstatic Truss Example 66
8.1.3 Post-Yield Behavior of the Hyperstatic Truss 68
8.1.4 Buckling of Compression Members in Hyperstatic Truss 70
8.1.5 Truss Deflections 72
8.2 Beam Failure 75
8.2.1 Beam Theory 76
8.2.2 Isostatic Beam Example 78

8.2.3 Hyperstatic Beam Example 80
8.2.4 Deflection Analysis 83
8.3 Creep Buckling Failure 86
8.3.1 Background 87
8.3.2 Restrained Column Model 87
8.3.3 Numerical Results 91
8.4 Discussion 93
References 94

Index 95

Chapter 1
Stability in Metallic Elements

Abstract The metallic alloys that are widely used in structures, machines and vehicles are chosen because they are, under many loading conditions, not subject to brittle fracture, and can withstand localized overloading. Nominally ductile metals can, however, fracture in a brittle manner under certain conditions. Cyclic loading can result in fatigue cracks that can lead to brittle fracture. Two modes of ductile instability are discussed: time independent loading and time dependent loading, or creep.

1.1 Time Independent Tensile Loading

The basic experiment for determining time independent loading properties in metals is the tensile test in which a prismatic specimen is subjected to unidirectional tensile loading. Its objective is to provide mechanical property data that can be used in the design of structures and machines. It provides, for example, data on the stiffness in the form of the elastic modulus, the transition from elastic to inelastic behavior, and the ultimate failure by fracture. Results from such tests are usually presented in the form of a stress-strain diagram such as shown in Fig. 1.1. By convention, the strain is plotted on the horizontal axis while the stress is plotted on the vertical axis, although in most testing the load (stress) is the independent variable.

Brittle metals exhibit only an initial elastic behavior leading directly to fracture. The behavior of a linearly elastic brittle material is characterized by the slope of the stress-strain which defines the elastic modulus and the fracture stress. Ductile metals such as low carbon steel exhibit a yield point, usually defined by the yield stress, which is the transition point between initial elastic and subsequent inelastic (or plastic) behavior (see Fig. 1.1). Detecting this transition can be difficult, and it is dependent on the sensitivity of the measuring instruments that are used. Since inelastic behavior is defined by the development of residual strain, it can be necessary to repeatedly apply increasing load/unload cycles until a residual strain is observed. Some materials such as steels exhibit a well-defined linearly elastic

R.L. Carlson et al., *Mechanics of Failure Mechanisms in Structures*, Solid Mechanics and Its Applications 187, DOI 10.1007/978-94-007-4252-9_1,

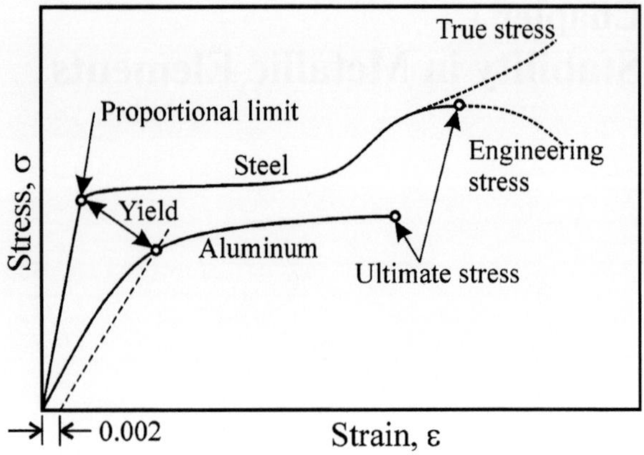

Fig. 1.1 Stress-strain diagrams for ductile and brittle metals

behavior up to the yield point followed by a nonlinear inelastic region until fracture at the ultimate stress as shown in Fig. 1.1. The transition point between linear and nonlinear behavior is called the proportional limit, and for steels it coincides with the yield point.

Ductile steels generally exhibit a characteristic post-yield behavior in which the material sustains increasing strain at an almost constant stress level until eventually the stress begins to increase and fracture finally occurs at what is defined as the ultimate stress.

Some metals, notably aluminum and copper, do not exhibit a well-defined initial linearly elastic region which makes it difficult to graphically determine the yield point. They also do not present an obvious proportional limit. An example stress-strain curve is also shown in Fig. 1.1. For practical purposes, the elastic modulus is measured as the initial slope, and experience has led to the adoption of the 0.2% offset yield stress as an estimation of the true yield stress. This is defined as the point of intersection between the stress-strain curve and an initial linearly elastic approximation offset by 0.2% strain. The material is assumed to behave linearly elastically up to the yield point so that the yield point and the proportional limit are the same (as they are for steels).

The strains developed for both time independent rupture and time dependent rupture are large, so the equations used for strain and stress for small strains must be replaced by equations for *true* strain and *true* stress.

The *engineering* strain equation is

$$\varepsilon_x = \left(L_f - L_0\right) / L_0 \tag{1.1}$$

where L_0 and L_f are the initial and final lengths. The *true* strain is expressed as the sum the increments of strain,

$$\varepsilon_x = \sum_{i=1}^{N} \frac{\Delta L_i}{L_i}. \tag{1.2}$$

By forming an integral and letting $\Delta L_i \rightarrow dL$, integration gives

$$\varepsilon_x = \ln\left(\frac{L_f}{L_0}\right). \tag{1.3}$$

It is often reasonable to assume that the material is incompressible so that the volume remains constant during the tensile test and therefore, $A_0 L_0 = A_f L_f$. Substituting this into Eq. 1.3 gives the following expression for the true strain

$$\varepsilon_x = \ln\left(\frac{A_0}{A_f}\right) = 2\ln\left(\frac{D_0}{D_f}\right) \tag{1.4}$$

where D_0 is the original diameter, and D_f is the final diameter of the local section.

The *engineering* stress is the current load divided by the original cross-sectional area. The definition of the *true* stress is the current load divided by the current area. If the material is incompressible, the increase in length of the specimen as load is applied will be accompanied by a decrease in the cross sectional area, and this will cause the true stress to be slightly greater than the engineering stress. The ultimate ductile failure by fracture is a form of tensile instability that is associated with the localized reduction in the cross sectional area called *necking* that leads to fracture. Figure 1.1 also shows the true stress curve for steel which is only slightly greater than the engineering stress until necking begins, and its sudden increase is due to the precipitous decrease in the cross sectional area in the necked region.

The value of the stress leading to fracture is described as the ultimate stress or the material tensile strength. This is usually recorded as the maximum engineering stress and occurs at fracture in tests when the load is controlled during the test. If the test is carried out under controlled strain, it is possible to record the downward turning dashed portion of the engineering stress curve in Fig. 1.1.

Under what are generally called service conditions for which a structure is expected to withstand repeated loads without permanent deformation, the design loads are determined by the yield stress for the material. If uncertainty is represented by a safety factor, N, then the maximum allowable stress is simply $\sigma_A = \sigma_Y / N$. Ductile metals are often used when overloads must be accommodated without catastrophic failure. In this case a limit load is defined based on a maximum allowable inelastic strain. For example, an automobile wheel suspension system would be designed to an allowable stress level for normal service conditions, and a limit load would be determined to prevent suspension collapse under the most extreme conditions.

When a bar is exposed to tensile loading in a structure, it usually is part of a statically indeterminate (hyperstatic) structural system. If additional loading beyond a maximum, elastic load for one element is applied, it will be transferred to adjacent, redundant elements. An effective analysis of statically indeterminate structures must, therefore, consider the load redistribution process. The analyses of these redistribution processes are presented in Chap. 8.

1.2 Tensile Creep Rupture

Hult (1974) has provided a comprehensive review of creep in continua and structures and has summarized failures due to several forms of damage. Hoff (1953) performed a tensile creep rupture analyses. His analysis was based on the use of constitutive equations that included only a time-dependent or creep component of strain. Subsequently, Carlson (1965) observed that the analysis of tensile creep rupture should also include a time independent, inelastic strain component, and presented an analysis in which an inelastic, time-independent component was included. This was consistent with the analysis of column creep buckling in which an inelastic time independent component is required. That is, the analyses of both column creep buckling and tensile creep rupture should include a time-independent strain component. Support for this conclusion is available in a report of the results of an analysis of tensile creep rupture tests on aluminum alloy specimens (Hoff 1953). The constitutive equation used in the analyses did not have a time-independent component. In all cases the predicted rupture times were greater than the test times.

Broberg (1974) and Böstrum (1975) have included time independent inelastic components in failure analyses of several structural systems exposed to creep environments.

1.3 Time Independent Compressive Loading

Time independent compressive loading rarely results directly in material failure in an engineered structure. Rather, compressive loading often leads to unstable equilibrium and arbitrarily large deflections lateral to the loading direction. This behavior is generally called buckling and can occur in axially loaded beams and columns as well as in plates and shells under in-plane loading. The fundamental analysis of time independent compressive instability or buckling is based on the use of a model for a perfectly straight, axially loaded column that is undergoing a transition from a straight to a bent form when a compressive load just sufficient to maintain the bent configuration is applied. The governing differential equation is

$$\frac{d^4w}{dx^4} + k\frac{d^2w}{dx^2} = 0 \tag{1.5}$$

in which $w(x)$ is the lateral deflection, $k = P/EI$, P is the applied axial force, E is Young's modulus, and I is the second area moment of the column cross-section about the axis of bending. The features of the buckling phenomenon can be considered by using the case of simply supported ends for which the deflections and the bending moments are zero at the ends of the column.

Within the elastic range, the buckling load, or Euler load, for simply supported (hinged) ends is

$$P_c = \pi^2 \frac{EI}{L^2}. \tag{1.6}$$

Dym (1974) has obtained a nonlinear solution for post-buckled behavior. The perfect column in the post-buckled regime is referred to as the *elastica*. The results do not apply for real columns.

The Euler load, P_c, obtained from Eq. 1.6 applies for elastic, perfectly straight, axially loaded columns. A factor responsible for the behavior observed for real columns is the fact that columns are neither perfectly straight nor axially loaded. The resulting imperfections cause columns to begin to deflect as soon as a load is applied. Also, the transition from an elastic to an inelastic behavior begins when the material on the concave side of the column begins to exceed the elastic limit. This varies from one material to another. It is of interest to observe, however, that in practice the alloys and the heat treatments chosen represent optimum strength conditions. As a result, the maximum loads obtainable are often close to the Euler load.

It is appropriate to observe that the type of column behavior described is obtained on statically determinant (isostatic) test systems. Most structural systems are statically indeterminate (hyperstatic), so load redistributions can occur that prevent a complete collapse of a structural system. This type of behavior is discussed in Chap. 8.

Two inspection procedures have been proposed for the inelastic buckling of short, axially loaded, straight columns. In the Reduced Modulus procedure proposed by von Kármán (1908) the transition from a straight to a bent form takes place under a constant load.

In the Tangent Modulus procedure proposed by Shanley (1947) the transition from a straight to a bent form takes place under an increasing load.

Test data from short, inelastic buckling tests are well correlated when tangent modulus values, E_T, replace E in Eq. 1.5. Tangent Modulus values can be obtained from material stress-strain curves.

1.4 Time Dependent Compressive Loading

In terms of the perspective developed here, time dependent buckling due to creep is an event that occurs at a critical time, or what might be described as the end of the column lifetime. The properties of the column material prior to buckling are of interest in this development, and concern is not just with creep buckling, but with the mechanics of a column for which deflections are increasing with time due to creep. Hoff (1954) has provided a criterion for column creep buckling. It is important to note that, as with tensile creep rupture considered by Carlson (1965) the constitutive equation used in the analyses must include both the time independent and time dependent components. An analysis by Fraeijs de Veubeke (1958) confirmed the earlier work by Hoff (1956). Knowing the critical deflection, the critical time must be computed by the use of the basic equation governing the problem. Fraeijs de Veubeke found that the critical deflection could be obtained by inspecting the column for stability at a deflection, v, by the consideration of a neighboring configuration, $v + \delta v$. Carlson and Breindel (1960) used the Hoff column model to focus on an inspection procedure that illustrated the abrupt transition from stable to unstable growth. When there is no creep, the loading curve rises to a maximum at which point buckling will occur. This represents *time independent* buckling behavior. If loading stops at a value that is less than the maximum load, and creep is present, the deflection will increase under a constant load. If sudden, virtual, compressive loads are introduced, the current stability state will be revealed by their load-deflection slopes. That is, buckling will be indicated when the *inspection load* slopes become zero.

In the analysis of this inspection procedure Carlson and Breindel (1960) differentiated the midpoint deflection function with respect to time and obtained a criterion for the *critical* deflection by setting the result equal to zero. The equation for the critical deflection coincided with that proposed by Fraeijs de Veubeke (1958).

In the discussion of time independent buckling it was observed that a column that buckled in a statically indeterminate system could be expected to transfer increases in loading to adjacent structural elements. In time dependent column creep buckling in a statically indeterminate system adjacent members can also assume additional loading. These elements may, in time, however, also become subject to creep buckling under increased loading. A time dependent, progressive sequence of individual column collapses may then occur. This may ultimately result in the total collapse of the structural system. A model for this behavior is presented in an example in Chap. 8.

1.5 Static Buckling Experiments

Elastic buckling can occur in a variety of configurations of bars and plates. Although analytical solutions are available for many of the problems of interest, the desirability of having a general experimental procedure for determining buckling loads was recognized, and Southwell (1932) proposed a procedure – the Southwell plot – that attracted considerable attention. Singer et al. (1998) have presented a comprehensive review of applications of the method proposed by Southwell.

They used the governing differential equation for a compressively loaded column to derive the Southwell plot relationship. The governing differential equation is

$$\frac{d^4w}{dx^4} + \alpha^2 \frac{d^2w}{dx^2} = -\alpha^2 \frac{d^2w_0}{dx^2} \tag{1.7}$$

in which $w(x)$ is the additional deflection due the loading, $w_O(x)$ is the initial, unloaded deflection, and

$$\alpha^2 = \frac{P}{EI}. \tag{1.8}$$

The boundary conditions for simple or hinged end supports are

$$0 = w(0) = \frac{d^2w(0)}{dx^2} = w(L) = \frac{d^2w(L)}{dx^2}. \tag{1.9}$$

The additional deflection, $w(x)$, due to the applied load can be represented by use of the Fourier series

$$w(x) = \sum_{n=1}^{\infty} W_n \sin\left(\frac{n\pi x}{L}\right). \tag{1.10}$$

The initial or unloaded deflection, $w_o(x)$, can also be represented by a Fourier series

$$w_0(x) = \sum_{n=1}^{\infty} W_{0n} \sin\left(\frac{n\pi x}{L}\right). \tag{1.11}$$

Equations 1.9, 1.10 and 1.11 substituted into Eq. 1.7 lead to the solution

$$W_n = W_{0n}\frac{1}{n^2\frac{P_E}{P} - 1}. \tag{1.12}$$

It then follows that

$$w(x) = \sum_{n=1}^{\infty} \frac{W_{0n}}{n^2\frac{P_E}{P} - 1} \sin\left(\frac{n\pi x}{L}\right) \tag{1.13}$$

in which $P_E = \pi^2 EI/L^2$ is the Euler load. The maximum deflection occurs at the midpoint so that

$$w(L/2) = w_{max} = W_1 - W_3 + W_5 - \ldots \tag{1.14}$$

where W_n is given by Eq. 1.12.

Southwell (1932) observed that as the buckling load is approached, that is when P becomes a large fraction of P_E, the fundamental or first term in Eq. 1.13 dominates. It then follows that $w = W_{01}/(P_E/P - 1)$, or after rearrangement

$$w = P_E\frac{w}{P} - W_{01} \tag{1.15}$$

From Eq. 1.15 it follows that slope of a plot of w/P versus w gives the buckling load, P_E, of the perfect version of the tested column. The intercept on the w axis gives the value of imperfection, W_{01}.

The initial data points on a Southwell plot often deviate from a straight line, because at low values of P, the higher deflection modes contribute to the measured deflection, and add to the total deflection. Lundquist (1938) recognized this and proposed a modification of the Southwell plot. His modification eliminated some initial deflection data points and used differences between an early, nonzero deflection value, δ_0, and subsequent values as measures of deflection. The results are then plots of $(\delta - \delta_0) / (P - P_0)$ versus $(\delta - \delta_0)$. The inverse of the slope then yields the buckling load, P_E.

The Southwell and Lundquist plots have been used for determining the buckling forces of a number of structural elements. Singer et al. (1998) have presented a thorough review of these applications.

Donnell (1938) considered the use of the Southwell plots for plates. On the basis of test results and an analysis of two plate problems, he concluded that to be valid, the imperfections and deflections must be small, and that the bent surface should very nearly be a developable surface. In these cases the compressive stresses were dominant and uniform.

An example of a plate problem in which compressive stresses are localized involves a tensioned plate with a slot with rounded ends. During tensile loading, the bottom and top edges of the slot develop compressive stresses. These edges will buckle or pucker when a tensile buckling load is achieved, and this behavior has been described by Carlson (1974). Buckling tests on a tensioned plate with a slot have been presented by Zielsdorff and Carlson (1972) and summarized by Singer et al. (1998). It was observed that a deviation from the linearity of a Lundquist plot of the test data indicated that middle surface stretching had developed. It was concluded that only the prior data should be included in the Lundquist plot.

To reveal the effects of the slot lengths and the radii at the ends of the slots, tests over a range of these values were conducted. It was found that the buckling load decreased with increasing slot length, and increased as the radius increased.

The results presented have a bearing on the potential effects on the plane stress fracture problem. Also, the effects of dynamic loading can be expected to further complicate the nature of the behavior. The results of an investigation of the parametric excitation are presented in Sect. 1.7.

1.6 Three Dimensional Problems

The previous sections have treated only uniaxial stresses, but engineered structures usually involve more complex stress and strain states up to and including fully three-dimensional states with significant values of all six components of stress and six of strain present. In this case, the state of stress at a point can be characterized by the three principal stresses and their orientation. Experiments have shown that yielding in ductile metals is independent of hydrostatic stress (the largest equal component of the principal stress state). Two yield criteria are commonly used: Tresca's and von Mises' (Bauchau and Craig 2009). The von Mises criterion (von Mises 1913) has several interpretations but the most physical is that yielding occurs when the maximum octahedral shear stress reaches a critical value. The octahedral shear stress is the shear stress on a plane making equal angles with the principal stress directions. Tresca's criterion is often called the maximum shear stress criterion and is based on the largest absolute difference between principal stresses. The two criteria generally bracket the experimental data, and von Mises' criterion is often preferred for its computational simplicity.

These criteria do not apply to brittle metals which do not yield but instead fracture. In most cases, failure of brittle metals depends on the maximum tensile stress developed within the material, and this can be determined from the largest principal stress.

1.7 Dynamic Stability

1.7.1 A Plate Problem

The loading conditions in the structural problems considered in previous sections didn't involve the inertial effects that are developed in vibratory motion. Bolotin (1964) has provided an introduction to the theory of dynamic stability and Simitses (1987) has written a review of the subject.

Experimental results described in Sect. 1.5 illustrated how static buckling failure can occur. Two examples in which dynamic instabilities can occur are considered in this Section. One involves a tensioned plate with a slot similar to the one discussed in Sect. 1.5. The second involves an eccentrically loaded bar under tensile loading.

The features of the plate problem are of special interest. The stress state prior to buckling is not uniform. The sheet can exhibit buckling both when tensioned and when compressed. That is, there are positive and negative eigenvalues.

To measure the deflection during testing, back-to-back strain gages were bonded on opposite faces of the plate along the edge of the slot. A record illustrating the development of large amplitude oscillations on the boundary of a principal region is shown in Fig. 1.2.

The test conditions were $P_{max} = 50$ lbs and $P_{min} = 8$ lbs. The recording paper speed was 100 mm/s. The record shown in Fig.1.1 represents the response observed in changing the excitation frequency from 41 to 42 Hz. The upper record represents the strain difference signal from the gages on the specimen surface. The lower record represents the load record. The buildup of the amplitude of the oscillation in the upper record is typical of the response observed in parametric excitation experiments.

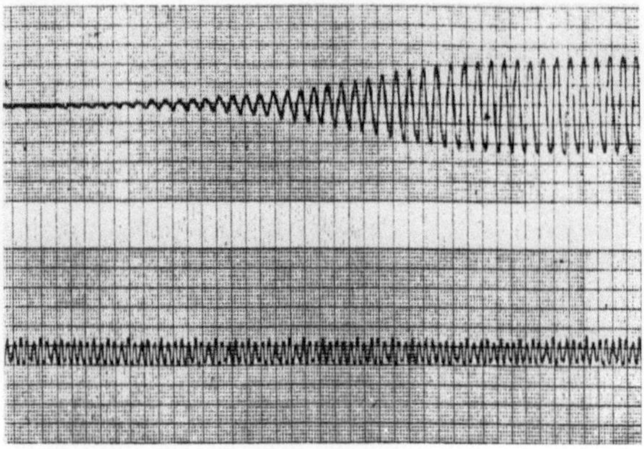

Fig. 1.2 Dynamic stability test records

1.7.2 A Tensioned Bar Under Eccentric Loading

An investigation of the parametric excitation of an eccentrically tensioned bar by Briley and Carlson (1977) revealed that two types of excursions are possible. Small excursions about the equilibrium were observed. These were on the same side of a line between the end loading points. Very large deflections were developed for larger alternating loads. This revealed the development of a 'snap- through' behavior that could result in oscillation amplitudes that are unacceptable.

References

Bauchau OA, Craig JI (2009) Structural analysis with applications to aerospace engineering. Springer, Dordrecht/New York

Bolotin VV (1964) The dynamic stability of elastic systems. Holden-Day, San Francisco

Böstrum PO (1975) Creep buckling considering material damage. Int J Solid Mech 11:765–775

Briley RP, Carlson RL (1977) An investigation of the parametric excitation of an eccentrically tensioned bar. Exp Mechanics 17(9):354–358

Broberg AA (1974) New criterion for brittle creep rupture. J Appl Mech 41:809–811

Carlson RL (1965) Creep induced tensile instability. J Mech Eng Sci 7:228–229

Carlson RL (1974) An experimental study of the parametric excitation of a tensioned sheet with a crack-like opening. Exp Mech 14(11):452–458

Carlson RL, Breindel WW (1960) On the mechanics of column creep. In: Hoff NJ (ed) Creep in structures. IUTAM Colloquium, Springer, pp 232–290

De Veubeke F (1958) Creep buckling. In: Hoff NJ (ed) High temperature effects in aircraft structures. Pergamon, London

Donnell LH (1938) On the application of Southwell's method for an analysis of buckling tests. In: The Stephen Timoshenko 60th Anniversary Volume. McGraw-Hill, New York, pp 27–38

Dym CL (1974) Stability theory and its applications to structural mechanics. Noordorff International Publishing, Leyden

Hoff NJ (1953) Necking and rupture of rods under tensile loads. J Appl Mech 20:105–108

Hoff NJ (1954) Buckling and stability. J Roy Aeronaut Soc 58:3–52

Hult J (1974) Creep in continua and structures. In: Zeman JL, Ziegler F (eds) Topics in applied mechanics. Springer, New York, pp 37–155
Lundquist EE (1938) Generalized analysis of experimental observation in problems of elastic stability. NACA TN 658
Shanley FR (1947) Inelastic column theory. J Aero Sci 14:261–268
Simitses GJ (1987) Instability of dynamically loaded structures. Appl Mech Rev 40(10): 1403–1408
Singer J, Arbocz J, Weller T (1998) Buckling experiments: experimental methods in buckling of thin-walled structures. Wiley, Chichester, pp 194–195
Southwell RV (1932) On the analysis of observations in problems of elastic stability. Proc Royal Soc, Series A 135:601–616, London
Von Kármán T (1908) Die knickfestigkeit gerader stäbe. Physikalische Zeitschrift 9(4):136–140
Von Mises R (1913) Mechanik der festen Körper im plastisch deformablen Zustand. Göttin Nachr Math Phys 1:582–592
Zielsdorff GF, Carlson RL (1972) On the buckling of tensioned sheets with cracks and slots. Eng Fract Mech 4:939–950

Chapter 2
Fracture Mechanics

Abstract The properties of the fracture of brittle materials has been recognized for many years. The brittle fracture of nominally ductile materials was recognized but not understood in the 1940s when nominally ductile metals were observed to fracture in a brittle manner. The evolution of an understanding of this behavior was developed in the early 1950s. The energy absorbed during the extension of a crack in a brittle solid is that which is associated with the surface tension of the increment of the crack extension. This result can be generalized to include brittle fracture in nominally ductile metals. Testing standards make it possible to determine fracture toughness values by the use of standardized test specimens.

2.1 Introduction

In an article Griffith (1920) performed an analysis in which he determined the energy that was absorbed during the extension of a crack in a brittle solid. He identified the absorbed energy as that which was associated with the surface tension of the increment of the crack extension. The results of his analysis were subsequently generalized to include brittle fracture in nominally ductile metals. Additional contributions by Irwin (1957) and Orowan (1952) ultimately led to the development of ASTM testing standards that make it possible to determine fracture toughness values by the use of standard test specimens.

Fracture toughness is a measure of how much stress is necessary to propagate a preexisting crack in a material, and since the presence of flaws in a material is unavoidable, this is an important material property. Flaws such as voids and inclusions, for example from a weld defect, or material discontinuities can give rise to cracks. Since flaws are essentially unavoidable, it is often necessary to assume a crack of some given size will be present in the material. The crack size and shape, the part geometry and loading, along with the material fracture toughness can be used to determine the ability of the part to resist fracture.

To be useful in fracture problems, it was necessary to develop relationships for crack energy absorption for specific crack geometries. Williams (1957) introduced

R.L. Carlson et al., *Mechanics of Failure Mechanisms in Structures*, Solid Mechanics and Its Applications 187, DOI 10.1007/978-94-007-4252-9_2,

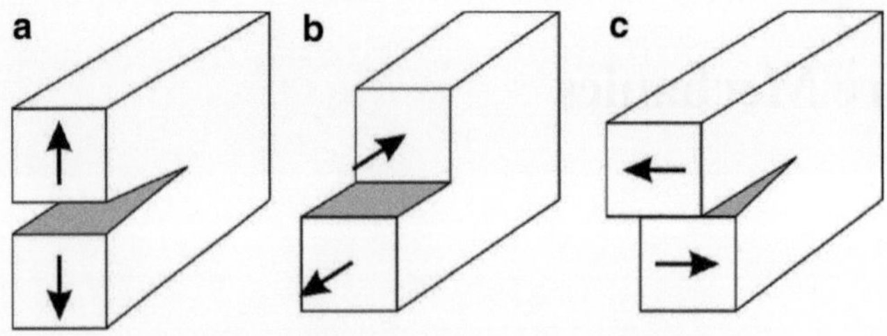

Fig. 2.1 Crack opening modes: (**a**) Mode I, (**b**) Mode II, (**c**) Mode III

an analysis in which the first term of a series expansion of a solution function provided the strength of the singularity of the first term. This term was defined as the stress intensity factor. The form of the stress intensity factor depends on the crack geometry and the details of the loading conditions. Crack geometries are defined as mode I (opening mode), mode II (sliding mode), and mode III (tearing mode) as illustrated in Fig. 2.1.

A fundamental relationship between the critical energy release rate for the crack opening (mode 1) fracture mode, G_{IC}, and the critical stress intensity factor, K_{IC}, provides a basis for the application of analyses of fracture in structural elements. For a mode I crack in plane strain crack this relationship has the form

$$G_{IC} = \frac{K^2{}_{IC}}{E}\left(1 - \nu^2\right) \tag{2.1}$$

where K_{IC} is the mode I critical stress intensity factor and E is the modulus of elasticity.

Applications in fracture mechanics require the identification of the geometry of crack that is of interest. This is required to determine the associated stress intensity factor K_I, which can then be used with the fracture critical energy release rate, G_{IC}, to determine the critical stress intensity factor, K_{IC}. Suresh (1991) has provided a detailed presentation of the experimental procedures and the analyses used to obtain values of G and their relations to critical stress intensity factors. Compilations of stress intensity factor values, K_{IC}, K_{IIC} and K_{IIIC} for structural alloys are available in the literature and in Ashby and Jones (1980).

If mode I fracture occurs in a brittle manner, the values of the stress intensity factor and the energy release rate are described as having reached their critical values. These are denoted as K_{IC} and G_{IC}, respectively. Fracture toughness, K_{IC}, is a mechanical property. Values for various alloys have been determined and are tabulated in the literature (Ashby and Jones 1980). Tests for determining these values are subject to restrictions which ensure that small scale yielding has occurred at the crack tip.

The application of the stress intensity factor, K, and its use in fatigue crack growth is considered in Chap. 3. The two sections that follow discuss cases in which applications of fracture mechanics are complicated by special metallurgical treatments or exposure time histories.

2.2 Fracture in Metals with Toughness Gradients

Equation 2.1 can be obtained by an examination of a procedure in which a crack of length a + α is reduced by an amount α, and the change in energy is calculated. This computation can be accomplished by use of the relation

$$G(\alpha) = \int_0^\alpha \sigma(r)u(\alpha - r)dr \tag{2.2}$$

where u is the crack face displacement between a and $a + \alpha$, and σ is the stress required to close this crack segment. The equality of Eq. 2.1 can be obtained from Eq. 2.2 by use of a Williams series expansions (Williams 1957) for σ and for u for the limiting condition in which α goes to 0.

In the analysis described above the energy release rate G is assumed to be constant. In some instances, however, microstructure variations with position and time can be developed. It has been observed by Carlson (1989) that G may vary with position. Examples include the base metal-weld zone regions cited by Bernard et al. (1982), steels in which hardenability affects the depth of hardening, and metals exposed to irradiation damage (Marschall and Rosenfield 1982).

If G in the material in front of a crack varies with position, it could be analytically represented as

$$G = G_0\left[1 + d_1\frac{r}{a} + d_2\left(\frac{r}{a}\right)^2\right] \tag{2.3}$$

where G_o, d_1 and d_2 are constants. Equation 2.1 should then be rewritten to account for gradient effects on G. Carlson and Saxena (1988) have observed that the sign of d_1 determines the type of loading response that will be developed.

When fracture toughness gradients are present, it may be anticipated that macro-residual stresses are present. For weld zones and simulated heat affected zones Braid and Knott (1982) concluded that fatigue crack propagation behaviors are significantly affected by the presence of residual stresses.

2.3 Metallurgical Embrittlement Effects

Metallurgical transformations in alloys that are exposed to elevated temperatures can result in significant reductions in strength. Aluminum alloys, for example, are not used in elevated temperature applications. This problem can, however, be avoided by selecting alloys that are not susceptible to reductions in strength within the operating temperature range for a given application. Creep induced brittleness has, however, been found to be a problem in alloys nominally considered to be acceptable for elevated temperature applications. The onset of brittle behavior was

Table 2.1 J_{IC} for Timken alloy 17-22-AS showing reduction in fracture toughness with time

Creep time, h	J_{IC}, kJ/m^2
0	71.8
6	54.3
42	13.6
100	13.3

a topic of an ASTM Symposium on Strength and Ductility of Metals at Elevated Temperatures (1952). A proposed experimental measure of the onset of brittle behavior consisted of a comparing results from creep rupture tests on notched and un-notched specimens. Although results obtained from tests from notched and un-notched specimens provide a good indication of the onset of brittle behavior, they do not provide data that can be used in design analyses. Subsequent investigations by Garofalo (1965) and by Siegfried (1971) focused on metallurgical reactions that introduce brittle behavior.

These approaches have contributed insights into phases of brittle behavior caused by creep. They have not, however, considered how the presence of a crack, for example, could affect the behavior of a structural element. The tolerance of an element to such a crack has been measured, and expressed in terms of fracture toughness by Carlson and Lo (1982). The Timken alloy 17-22-AS was chosen for the tests, because a previous experimental investigation by Jones et al. (1955) provided basic data from notched and un-notched tensile tests on this alloy. Interrupted creep tests at 1,100°F were conducted on tensile specimens for 6, 42, and 100 h. The objective was to provide a comparison the fracture toughness values after increasing creep exposure times. After creep testing, a notched bend specimen was machined in each of the creep test specimens. Each of the bend specimens were then fatigue tested with three point bending to produce cracks at the notch tips. These tests were conducted at the same temperature used for the creep tests; that is, 1,100°F.

Experiments for obtaining fracture toughness values, J_{IC}, were conducted on the cracked specimens using a procedure proposed by Rice, Paris and Merkle (Rice et al. 1973), and subsequently developed by Landes and Begley (1976). The procedure adopted consisted of loading the specimen in three point bending. The deflection at the mid-point was measured by a clip gage outside of the furnace. The tests were interrupted at 15 different deflection values to produce increments of crack extension. After testing, the specimens were removed from the furnace, and broken apart. The machined, the fatigue and the crack extension surfaces easily discernable because of the differences in surface textures, and the degree of oxidation. The crack extension lengths were measured with an optical comparator. The areas under the load-deflection curves were measured by use of a planimeter. J integral values computed from the test data measurements were plotted versus the crack extension values, and the results were used to determine J_{IC} values for each of the tensile creep test times.

Values of J_{IC} for the four creep exposure times are presented in Table 2.1. Fracture toughness has decreased substantially with time.

References

Ashby MF, Jones DRH (1980) Engineering materials 1: an introduction to their properties and applications. Pergamon, Oxford

Bernard G, Devillers L, Faure F, Marandet B (1982) In: Francois D (ed) Advances in fracture research, vol 3. Pergamon, Oxford, pp 1227–1234

Braid JEM, Knott JF (1982) Advances in fracture research, vol 5. Pergamon, Oxford, pp 2061–2069

Carlson RL (1989) Crack extension in metals with fracture toughness gradients. Int J Fract 41(3):R51–R54

Carlson RL, Lo HC (1982) A fracture mechanics study of creep induced embritlement. Int J Fract 18:145–157

Carlson RL, Saxena A (1988) On crack extension sensitivity. Int J Fract 37:101–106

Garofalo F (1965) Fundamentals of creep and creep rupture in metals. Macmillan, New York, pp 213–241

Griffith AA (1920) The phenomena of rupture and flow in solids. Phil Trans Royal Soc 221A: 163–198

Irwin GR (1957) Analysis of stresses and strains near the end of a crack traversing a plate. J Appl Mech 24:361–364

Jones MH, Newman DP, Sachs G, Brown WF (1955) Effects of variation in normalizing and tempering procedure on stress rupture strength, creep embrittlement and notch sensitivity for a cr-mo-v and a 17 cr-4ni-cu steel. Trans ASM 47:926–954

Landes JD, Begley JA (1976) A fracture mechanics approach to creep crack growth. In: Rice JR, Paris PC (eds) Mechanics of crack growth, vol 590, ASTM STP. American Society for Testing and Materials, Philadelphia, pp 170–186

Marschall CW, Rosenfield AR (1982) Advances in fracture research, vol 5. Pergamon, Oxford, pp 2467–2475

Orowan E (1952) In: Murray WM (ed) Fundamentals of brittle behavior in metals in fatigue and fracture of metals. Wiley, New York, pp 139–167

Rice JR, Paris PC, Merkle JG (1973) Some further results on J-integral analysis and estimates. In: Kaufmann JG (ed) Progress in flaw growth and fracture toughness, vol 536, ASTM Special Technical Publication. American Society for Testing and Materials, Philadelphia, pp 231–245

Siegfried W (1971) In: Nicholson AM (ed) Advances in creep design. Applied Science Publishers, London, pp 181–222

Suresh S (1991) Fatigue of materials. Cambridge University Press, Cambridge, pp 141–189

Symposium on strength and ductility of metals at elevated temperatures (1952) ASTM Special Technical Publication 128, American Society for Testing Materials, Philadelphia

Williams ML (1957) On the stress distribution at the base of a stationary crack. J Appl Mech 24:109–114

Chapter 3
Fatigue in Metals

Abstract Fatigue, as a mechanism of failure in metals, has been of interest for many years. Its importance is indicated by an observation by Pook (Pook L (1983) The role of crack growth in metal fatigue. The Metals Society, London) who indicated that between 1938 and 1983 about 20,000 papers were published on the subject of fatigue. The initial emphasis on fatigue focused on the formation of cracks on smooth surfaces after prolonged periods of cyclic loading. In early work an emphasis was focused on determining the number of cycles required to cause failure by fracture. As the details of fracture mechanics evolved, interest in the development of a crack growth relation resulted in the development of a new approach to the analysis of fatigue crack growth. The operating conditions of structural elements can vary widely. These include variable amplitude loading and environmental conditions. The latter includes temperature and exposure to corrosive atmospheres. The effects of these on fatigue crack growth should be included in structural analyses, and each are considered in the following sections.

3.1 Introduction

The term fatigue is not exclusively associated with cyclic loading in metals. Although the mechanisms differ, fatigue damage has also been found to occur in other solids. Details of these are discussed in Chaps. 4 through 6. To provide a perspective for the scope of the fatigue phenomena which have been identified, a description of the qualitative features of fatigue in both metals and nonmetals serves as an introduction to the more detailed presentations which are provided in subsequent chapters. This can be visualized by reference to Fig. 3.1 which is a plot of the stress range versus the log of the number of loading cycles.

In this representation the minimum stress on the specimen is zero so that the stress range is equal to the maximum stress. A specimen with a maximum stress value corresponding to B will have a crack initiation site for the number of loading

R.L. Carlson et al., *Mechanics of Failure Mechanisms in Structures*, Solid Mechanics and Its Applications 187, DOI 10.1007/978-94-007-4252-9_3,

Fig. 3.1 Fatigue crack initiation and failure curves

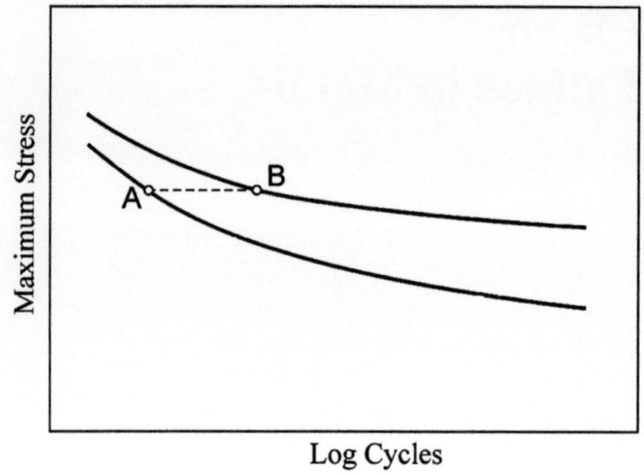

Fig. 3.2 Statistical representation of scatter in the number of cycles to failure

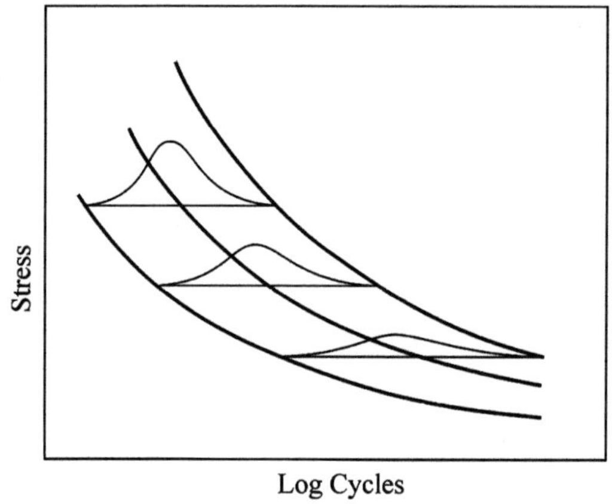

cycles at A. That is, failure by fracture is preceded by crack initiation. The failure time is given by the number of cycles at point B. The horizontal line AB represents all phases of the fatigue for the given maximum stress, crack nucleation, crack growth, and final failure by fracture. If a line from point A never intersects an upper curve, the initiated crack is described as being non-propagating or arrested.

A large portion of the total time to failure, as shown in Fig. 3.1, is consumed during the crack nucleation phase. Since the time of this phase can vary significantly from one test specimen to another, considerable data scatter can occur in the number cycles to failure tests. For a given stress level, it is necessary to consider the use of statistics to deduce the probability of failure at each stress level. This is qualitatively illustrated in Fig. 3.2.

Note that the scatter is largest in the low stress region where the stresses approach what is defined as the endurance limit. This is discussed in Sect. 3.2. The basic elements of the statistical analyses of fatigue data have been discussed by Weibull (1961) and Johnson (1864).

3.2 The Stress-Life Strategy

Evaluations of the features of the stress versus loading cycle representations have led to the development of the stress-life strategy. After analyses of test data, the features of stress versus log of the number of cycles curves have been found to have two distinct forms. The curves shown in Figs. 3.3 and 3.4 illustrate the features of two types of *stress-life curves*.

The curve in Fig. 3.3 decreases asymptotically to a horizontal, constant stress value. Stress amplitudes that are less than this value do not result in failure. Micro-cracks may develop below this value, but growth is arrested. The lower bound stress value is described as the endurance or fatigue limit. This limiting stress value is most commonly observed in low strength steels.

The curve in Fig. 3.4 represents cases for which there are no lower bounds or fatigue limits. Most nonferrous metals do not exhibit a lower bound stress.

For applications in which the amplitude of the cyclic loading is more or less uniformly repeatable and the desired lifetimes involve millions of cycles with moderate levels of loading, design can be based on data obtained from tests in which the loading is of the type depicted in Fig. 3.5.

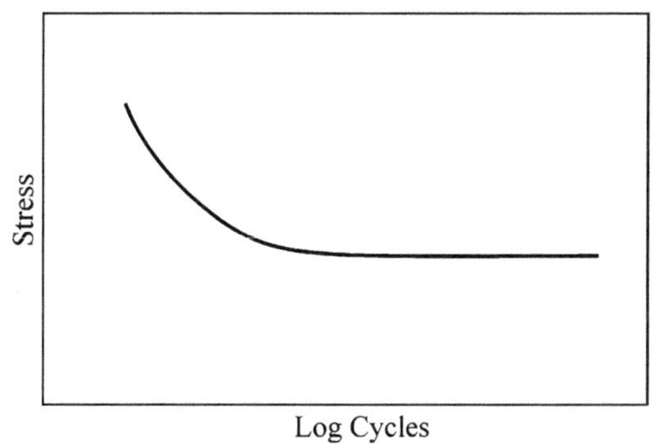

Fig. 3.3 Stress versus log of cycles diagram with fatigue limit

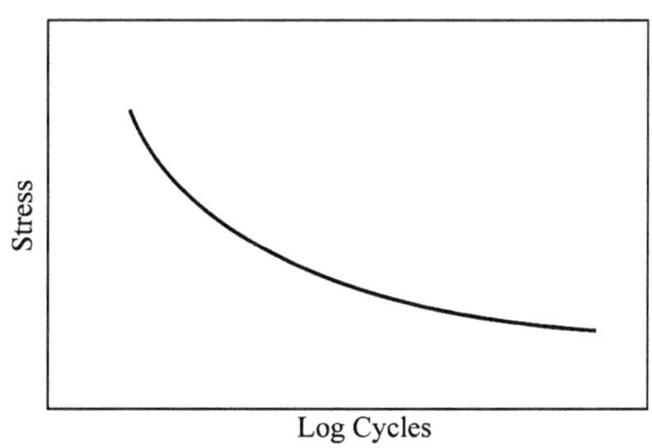

Fig. 3.4 Stress versus log of cycles diagram with no fatigue limit

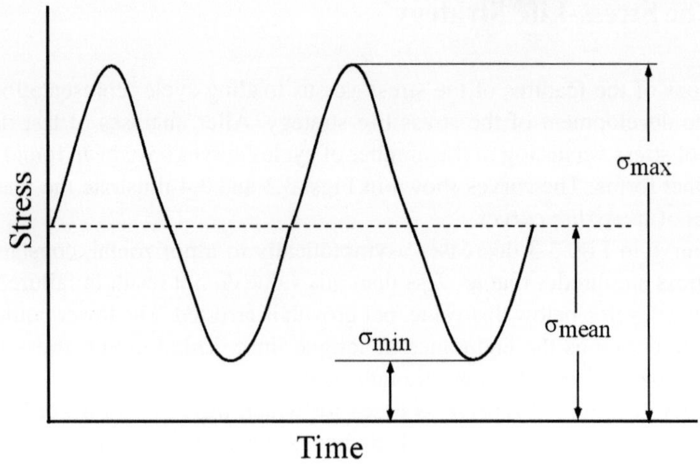

Fig. 3.5 Cyclic variation of stress

Two loading parameters which are used are, by definition, the stress range

$$\Delta\sigma = \sigma_{max} - \sigma_{min}, \tag{3.1}$$

and the stress ratio

$$R = \frac{\sigma_{min}}{\sigma_{max}}. \tag{3.2}$$

Since σ_{min} can be chosen to be compressive, it follows that R can be negative.

3.2.1 *The Stress-Life Diagram (S-N Curve)*

To obtain a stress-life diagram, uniform, constant amplitude tension-tension (positive stress ratio R) and tension-compression (negative R) stress-life tests are conducted. The test traditionally used to obtain basic stress-life data (also called S-N curve) is the rotating bend test for which the stress ratio $R = -1$. Figure 3.6 shows the schematic of a four-point rotating bend test. The applied load is static but the specimen is in rotation. Because of the rotation, the entire surface material is tested under maximum stress. A cylindrical, hourglass gage section which has a highly polished surface is used. The high rotational speeds which are employed make it possible to accumulate large number of cycles in a reasonably short period of time. The stress values used from reversed bending tests are the maximum bending stresses computed from elementary beam theory.

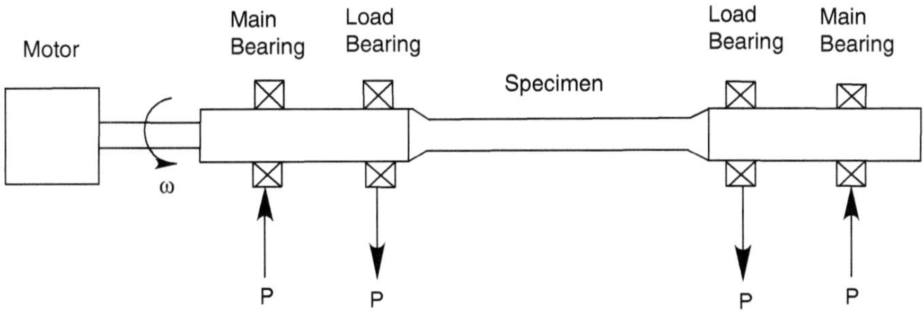

Fig. 3.6 Schematic of a four-point rotating bend test. The main features are: (**a**) the motor, which gives the rotation (**b**) the two main bearings (supports) and (**c**) the two load bearings (where the static load is applied)

Because of data scatter, multiple tests are conducted at each stress level, and the test data are summarized on stress- log cycles to failure plots, of the type shown in Figs. 3.3 and 3.4, which represent the two distinct behaviors which are encountered (fatigue with endurance limit or no fatigue limit).

3.2.2 Mean Stress Effects – The Goodman Relationship

The stress histories for many components do not have a mean stress of zero as the reversed loading, $R = -1$ case. Mean values can be either tensile or compressive. The nonzero mean can be due either to externally applied loading or to residual surface layers stresses. The parameters used to describe mean stress effects are defined in terms of the maximum and minimum cyclic stresses. These are the mean stress,

$$\sigma_m = \frac{1}{2}\left(\sigma_{\max} + \sigma_{\min}\right), \tag{3.3}$$

and the alternating stress,

$$\sigma_m = \frac{1}{2}\left(\sigma_{\max} - \sigma_{\min}\right). \tag{3.4}$$

These two quantities can be expressed by dimensionless ratios (σ_m/σ_f) where σ_f is the fatigue strength for reversed loading and as (σ_m/σ_y) where σ_y is the yield strength or as (σ_m/σ_u) where σ_u is the ultimate strength. Reference to the values in the denominators provides a basis for developing empirical relations for describing mean stress effects.

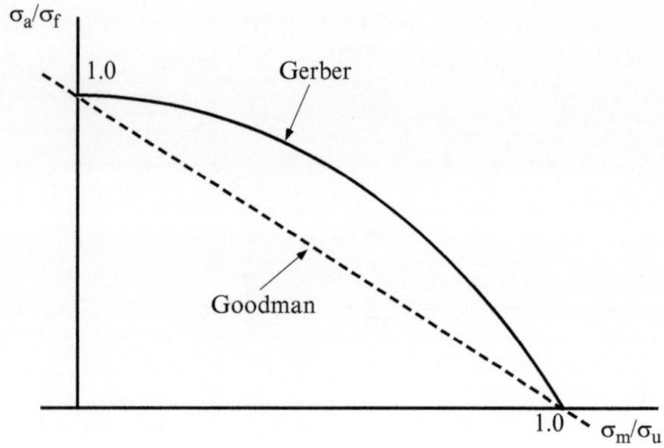

Fig. 3.7 The Goodman diagram (*dashed line*) and the Gerber diagram (*solid line*)

A common approach for developing empirical equations is to represent a dependent variable in terms of a power series in an independent variable. To correlate this type of approach with equations which have been widely used we select the ratio (σ_a/σ_f) as the dependent variable and either (σ_m/σ_y) or (σ_m/σ_u) as the independent variable. Thus, we can write, for example,

$$\frac{\sigma_a}{\sigma_f} = C_1 + C_2\left(\frac{\sigma_m}{\sigma_u}\right) + C_3\left(\frac{\sigma_m}{\sigma_u}\right)^2 + \cdots. \tag{3.5}$$

If three terms were to be used, three conditions for values of the ratio (σ_m/σ_u) would be required to evaluate the coefficients. Only two conditions are usually used, however. One procedure, which has been used, is to observe that for $\sigma_m = 0$ then $\sigma_m = \sigma_f$ for failure, hence $C_1 = 1$. Then, keeping only two terms in the series, and observing that for $\sigma_a = 0$, then $\sigma_m = \sigma_u$ for failure, we obtain $C_2 = C_1$. Thus, Eq. 3.5 becomes

$$\frac{\sigma_a}{\sigma_f} + \frac{\sigma_m}{\sigma_u} = 1. \tag{3.6}$$

This equation, proposed by Goodman (1899), is known as the Goodman equation. This linear relation is shown in Fig. 3.7.

Since test data tend to lie above the straight line, it follows that a nonlinear curve lying above might be preferable. Therefore, proceeding as before and keeping only the C_1 and C_3 terms in Eq. 3.5 results in the following equation, credited to Gerber (1874)

$$\frac{\sigma_a}{\sigma_f} + \left(\frac{\sigma_m}{\sigma_u}\right)^2 = 1. \tag{3.7}$$

This equation is also shown in Fig. 3.7.

3.2.3 Variable Amplitude Loading – The Palmgren-Miner Rule

Aerospace structures are subjected to varying amplitudes of loading with time rather than repetitive cyclic loading of constant amplitude. In fact, the loading spectra are not only complex, but differ for different types of aircraft, e.g. fighter aircraft, transport aircraft.

Proposals by Palmgren (1924) and Miner (1945) attempt to account for variable amplitude loading effects by the use of the assumption that the fatigue damage incurred during each cycle is independent of the prior loading history. The rule based on this assumption is described as the linear damage rule. If the amplitude of loading is σ_i for n_i cycles and the fatigue life at σ_i is N_i, the fraction of life used would then be n_i/N_i. Therefore, for a total of k cycles in a loading spectrum consisting of n_1 cycles at stress level σ_1, n_2 cycles at stress level σ_2, etc., fatigue failure occurs when

$$\sum_{i=1}^{N} \frac{n_i}{N_i} = 1. \tag{3.8}$$

Although the linear damage rule is intended to provide a basis for predicting fatigue life under variable amplitude loading, it has deficiencies which should be recognized. Experiments have shown that the sum of Eq. 3.8 can, depending upon the order in which load levels are applied, be either greater or less than unity. If, for example, a block of high level loading is followed by a block of low level loading, the sum for failure in Eq. 3.8 can be less than unity. For notched specimens the reverse, however, is found to occur. If a block order is changed to a low-high sequence, the sum for failure is greater than unity. It has been argued that if the loading spectrum is a mixture of high-low and low-high sequences, the deviations from unity may be canceled. Alternative methods for assessing the accumulation of damage under variable amplitude loading have been proposed and some of these are discussed by Collins (1981). The introduction of material constants and the complexity of applying them to complex spectra, however, often detract from their usefulness.

There are approaches to include other effects in the basic S-N approach, namely *multi-axial stress states* (e.g. combined torsion and bending), effects of *stress concentration locations* such as holes, notches and fillets and effects of the *environment* (extremes in temperature and active or corrosive atmospheres); these are discussed in detail in Carlson and Kardomateas (1996).

3.2.4 Fatigue Design Approaches: Safe-Life, Fail-Safe and Damage Tolerance

During the 1950s, the loss of several Comet aircraft within a short period of time led to an investigation into the structural integrity of Comet's fuselage. A pressurization

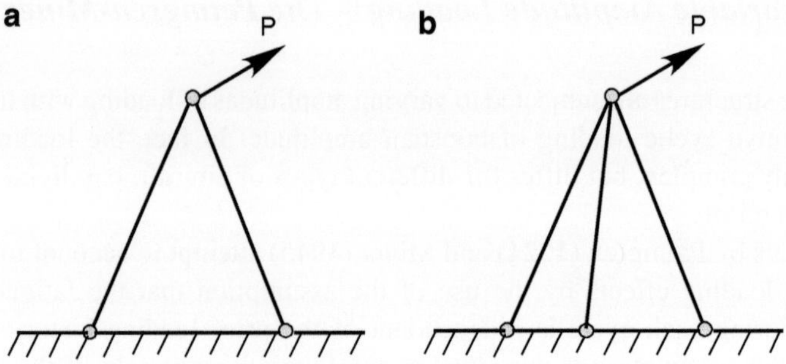

Fig. 3.8 Trusses: (**a**) A two-bar truss and (**b**) a three-bar truss (fail-safe)

test on a sample aircraft revealed that the combination of fatigue and stress concentration at the corner of a window was the probable cause of the failures. These accidents as well as others prompted the development of fatigue methods for aircraft design. The first and oldest approach to dealing with fatigue is to require that a structure should be able to survive several times the intended service lifetime, e.g. the time to failure of an aircraft wing in a laboratory fatigue test may be required to be four times the expected service lifetime. This type of requirement constitutes a design philosophy described as *safe-life*. The safe-life approach determines a replacement time for aircraft components, usually specified as a number of allowable landings or flight hours. Once a component reaches its replacement time, its safe-life is considered to be used up and it is retired, whether or not any fatigue cracks are present. There are, however, two significant problems with this method: (a) the safety of an aircraft is not protected if a manufacturing or accident or maintenance induced defect is introduced and (b) safe-life safety factors are quite conservative and thus many components can be quite prematurely retired.

The next approach used in aircraft fatigue design is the *fail-safe* approach, which was developed in the 1960s and is based on designing structures with multiple load paths, such that if an individual element fails, the remaining elements can carry the additional load from the failed element until the damage is detected and repaired. Indeed, there are designs in which the failure of a single component results in a total system failure. As a simple example, consider a two bar truss under an external load (Fig. 3.8a). If one bar fails, the remaining bar can no longer serve the intended structural function. By contrast, the failure of one bar in a three bar truss (Fig. 3.8b), simply results in a redistribution of the bar forces and the truss can still, at least temporarily, serve its intended function of the truss. The latter example can then be described as being fail-safe. This concept is discussed in more detail in Chap. 8.

Fail-safe design is obviously a desirable solution, and it is commonly used in aircraft. Ideally, an aircraft designed according to fail-safe principles can sustain damage and remain airworthy until the damage is detected and repaired. This philosophy necessitates periodic inspections to assess the integrity of the load carrying members. However, this philosophy cannot preclude the possibility of simultaneous crack development in multiple load path elements and the inspections in this approach were not based on crack growth principles (fracture mechanics); as a result, the loss of several *fail-safe* aircraft in the mid-1970s emphasized the need to locate cracks and repair them before failure occurred.

The presence of a crack in a component subject to load variations does not necessarily constitute failure. The crack may undergo a time-dependent extension, which is often described as *stable* growth. Eventually, the growing crack may attain what is described as a critical length and then unstable or catastrophic growth can occur. Thus, during a stable growth period, the structural integrity of the system remains intact and the primary concern is the anticipation of when a critical length will be attained. This requires knowledge of the loading history on the cracked component. Prescribed inspection intervals are instituted and fracture mechanics is employed to ensure that a crack would not grow to its critical length within the inspection interval time. Thus, cracks occurring at any time, would be caught at the next inspection interval before they have a chance to become critical. Naturally, associated repairs would ensue. Although the task described is complex and a simplified description has been presented, the logic involves evaluating the tolerance of components to the presence of cracks. This is the basis for the evolution of a sophisticated procedure, which is described as *damage tolerance*. In the early 1970s, the U.S. Air Force was the first to adopt the damage tolerance fatigue design approach. With economic and safety advantages over the previous methods, the damage tolerance philosophy was eventually adopted by the commercial aviation.

As mentioned, the objective of the damage tolerance approach is to detect cracks in principal structural elements before their critical length. A principal structural element (PSE) is defined as any aircraft structural component carrying flight, ground, or pressurization loads, whose failure could result in the loss of the aircraft. The goal is to establish inspection intervals for these elements based upon the time it takes to grow a crack from an initial detectable size to the critical crack length.

The first task in the aircraft damage tolerance approach is to define the usage profile. This profile describes the various flight conditions, such as taxi, climb, cruise, descent, and landing impact, and the amount of time spent at each gross weight, speed, and altitude. The usage profile is then used to create a load factor spectrum at the center of gravity of the aircraft. The next task is to identify the principal structural elements (PSE's), convert the load factor spectrum into a stress spectrum for each location and incorporate the effects of the service environment. Using crack growth (*da/dN)* equations, such as the Paris or Forman equations (described next), the stress spectrum is combined with material properties data and stress intensity factor solutions applicable to each PSE to determine the number of cycles for a crack to reach the critical length starting from the detectable length. This number is usually divided by a factor of 2 to arrive at the inspection interval. This ensures that, should a PSE develop a crack, it will be inspected at least once before the crack propagates to failure.

Unlike the safe-life approach where components are retired whether or not they are damaged, in the damage tolerance approach components are only replaced if a crack is found during an inspection. It is important to note that any size crack found during an inspection mandates replacement of the damaged component (even if the crack is at the just detectable size and therefore the component could last until the next inspection). Another advantage of the damage tolerance approach is that crack growth is rather deterministic unlike the large scatter associated with the S-N methodology. Thus, it allows for a reduction of safety factors in design.

3.3 Fatigue Crack Growth

The determination of how many cycles of loading can be applied before a critical condition is reached requires the availability of an equation for determining crack growth. Also, the methodology implies that the sites that have been identified as critical should be monitored by use of non-destructive testing instruments. The elements of the concepts outlined form the basis of a method that evaluates damage tolerance. Structural integrity can then be judged by the extent to which a tolerance to damage is achieved. In view of the scatter associated with the collection of data for the stress-life method, concern about comparable difficulties might be anticipated in an implementation of a crack growth strategy. It was, however, observed by Frost et al. (1974) that the total contribution to the scatter in the *long* crack growth regime is much less than that in the *small* crack initiation phase. The statistical details of this behavior are discussed in a section on *small* fatigue crack growth. The interest in the use of fracture mechanics in fatigue led to a search for a crack growth relationship that described the cyclic rate of crack growth as a function of stress and crack length. In functional form this results in an equation of the form

$$\frac{da}{dN} = F(\sigma, a) \tag{3.9}$$

in which *da/dN* is the cyclic rate of crack growth, σ is the stress, and a is the crack length. When a power function is chosen, F can be written as

$$F = \sigma^m a^n. \tag{3.10}$$

Frost (1959) proposed values of $m = 3$ and $n = 1$ that could be used to correlate laboratory results. Liu (1963) suggested that the range of stress, $\Delta\sigma$, should be substituted for σ, and proposed values of $m = 2$, and $n = 1$.

Paris et al. (1961) recognized that the mode 1 stress intensity factor includes the loading details and the geometric details of the geometries of the component. They, therefore, proposed a crack growth rate law that incorporated their ideas.

Donaldson and Anderson (1962) subsequently presented support for the use of the mode one stress intensity factor by demonstrating successful correlations for two aluminum alloys, and two steels.

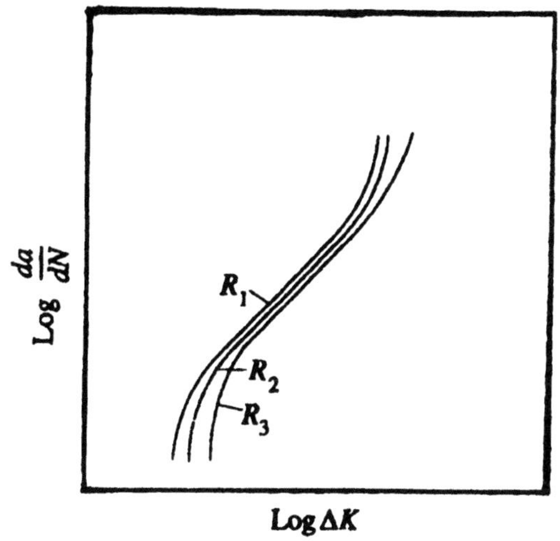

Fig. 3.9 Crack growth rate versus mode I range of stress intensity factor for $R_1 > R_2 > R_3$

Paris and Erdogan (1963) then suggested a crack growth rate law of the form

$$\frac{da}{dN} = C(\Delta K)^n \tag{3.11}$$

where ΔK is the range of the crack growth mode I stress intensity factor. An illustration of log *da/dN* versus log ΔK plots for different values of the ratio $R = \sigma_{min}/\sigma_{max}$ is presented in Fig. 3.9.

The linear, intermediate Paris region in Fig. 3.9 represents growth in the *long* crack regime. The final, accelerating portion of the curve represents the final phase leading to fracture. The region to the left of the Threshold is described as the small crack regime. Small cracks can be initiated, and they can grow within this region. Crack initiation and growth within this region are considered in Sect. 3.4.

The stress intensity factor makes it possible to use basic fatigue crack growth data for a variety of crack geometries and loading conditions. Crack growth rates for two different loading cases can be expected to be the same if, for example, the values of ΔK are the same. To consider the issues involved in this problem, it is necessary to review possible differences between such cases. This requires a consideration of the role of the stress ratio $R = \sigma_{min}/\sigma_{max}$. It is clear that two values of $\Delta\sigma$ may be equal, but can have different values of *R*. A plot of data with equal $\Delta\sigma$ values, but different values of *R* is presented in Fig. 3.9. It illustrates a type of behavior that has been observed. Distinct curves are observed for different values of the stress ratio, *R*. As shown in Fig. 3.9, the growth rates increase with increasing *R* for a given value of ΔK. The results shown in Fig. 3.9 indicate that ΔK should to be modified to account for the effect of the value of R on ΔK. This has been has been accomplished by the introduction of an equation of the form

$$\Delta K_{eff} = F(R)\,\Delta K. \tag{3.12}$$

A variety of equations have been proposed for F. These have the form

$$F(R) = \left(a_1 + a_2R + a_3R^2\right). \tag{3.13}$$

Values for the coefficients in Eq. 3.13 differ for each alloy. Values for the coefficients in Eq. 3.13 have been proposed by Forth et al. (2007) and by Forth et al. (2006).

3.4 Crack Initiation Mechanisms

Fatigue damage that results in the nucleation of cracks can be initiated in a layer on a free surface. They can occur as a result of localized, irreversible slip in narrow bands within individual grains. Since the band regions terminate at free, unconstrained surfaces, roughness in the form of extruded material can be created. High concentrations of plastic strain are developed within these localized bands, and they can ultimately serve as sites for the initiation of micro-cracks. For a site with a large stress concentration factor a long, single crack can develop, and its growth can be computed by use of the analysis described in the previous section.

The period of growth in the small crack regime can be a large portion of the total life. Also, multi-site cracks can be initiated at sites with small stress concentrations. The occurrence of multi-site cracking in the small crack regime, therefore, introduces a need for a statistical consideration of the growth behavior in the small crack regime. The statistical elements of a multi-site analysis are presented in the next section.

3.5 Multi-site Fatigue Cracking

In the preceding section the use of the effective stress intensity factor, ΔK_{eff}, was confined to applications involving a single crack. The analyses involved do not apply in situations in which multiple cracks are present. The network of cracks in this situation can be expected to interact with one another. Swain examined multi-site cracking, and suggested that since only one crack in a cluster of cracks results in failure by fracture, data from these should be described as *valid data*. The remaining data should be described as *invalid data*. In this presentation the *valid data* are described as *primary data* and the *invalid data* are described as *secondary data*. The networks of the secondary cracks, along with the grain boundaries, form the neighborhoods within which the primary cracks grow. A collection of all of the crack growth data from multiple specimens has a bimodal distribution.

The primary crack in a cluster of cracks in a specimen cannot be easily identified during the early crack initiation phase. The selection process requires the use of Swain's criterion (Swain 1992).

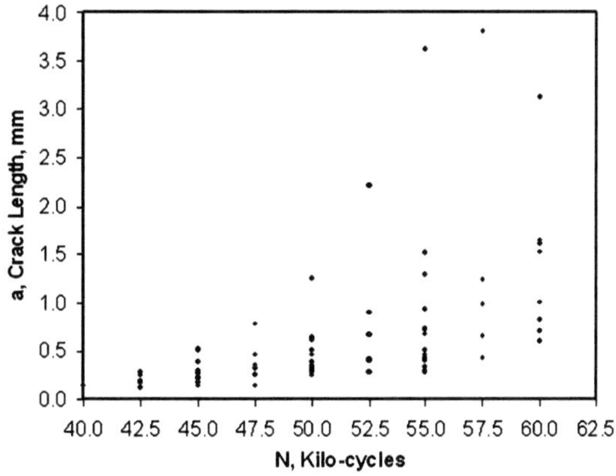

Fig. 3.10 Primary crack data

The crack growth histories of both the primary and secondary cracks must, therefore, be obtained for each specimen in a collection of specimens. The results of an experimental application of Swain's criterion are presented in the next section.

3.5.1 Multiple Site Cracking Test Results

Multi-site fatigue crack tests have been conducted by Carlson et al. (2007). A double edge notch test specimen of the aluminum alloy 7075-T7351 was used. The tensile stress at the bottoms of the notch surfaces was 1.2 times the average stress in the reduced section. This provided surfaces upon which multiple micro-cracks could be developed. These areas were polished with abrasive papers and diamond pastes to prepare the surfaces for crack length measurements. The experimental program was designed to trace and record the evolution of the crack cluster patterns on the bottoms of the notched surfaces. Crack lengths were measured by interrupting the loading at cyclic intervals. A tele-microscope was used to measure crack lengths. The crack lengths within each of the clusters were measured on both faces of the reduced sections.

Te objective of the procedure was to develop crack growth histories. By measuring all of the cracks for each cyclic load interruption, it was possible to select the emerging, primary crack. Once it was identified, its growth was traced back to measurements made during earlier load interruptions. A review of these measurement records then made it possible to identify the primary cracks and the secondary cracks. Crack measurement data were obtained on ten surfaces of five specimens, and a total of 55 cracks were measured. Of these, ten were identified as primary cracks. The remaining were secondary cracks. The primary crack data are presented in Fig. 3.10, and the secondary crack data are presented in Fig. 3.11.

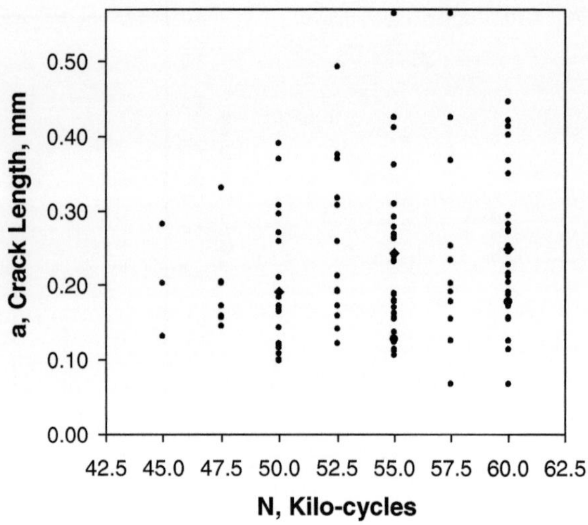

Fig. 3.11 Secondary crack data

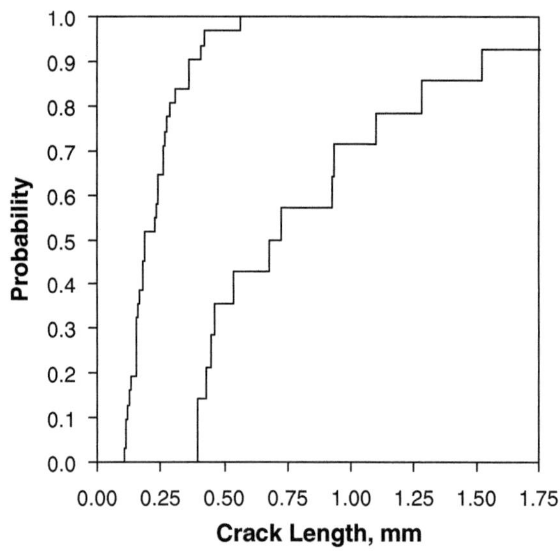

Fig. 3.12 Probabilities for secondary cracks (*left curve*) and primary cracks (*right curve*)

An examination of Figs. 3.10 and 3.11 indicates that the range of the secondary cracks is substantially smaller than that of the primary cracks. Also, the primary cracks exhibit increasing growth rates, whereas the secondary cracks indicate the onset of arrested growth.

The data for the primary and secondary cracks have been used to construct the cumulative frequency polygons (Hald 1957). These are presented in Fig. 3.12 for $N = 55{,}000$ cycles of loading. Each horizontal step represents the probability for one data value. As would be expected, there are more secondary steps than primary steps. Note also, that the largest secondary cracks are of the order of the smallest primary cracks.

Swain's observation that the total crack population of clusters of small fatigue cracks should be separated into two separate categories has been confirmed. This observation forms a basis for developing a bimodal representation of the multiple-site fatigue crack data. Analyses of scatter in multi-site cracking should include only the primary of *valid* data that ultimately lead to fracture.

3.5.2 Transition from Small to Large Fatigue Cracks

The effective stress intensity factors are not applicable in the small crack growth regime when multi-site cracking is present. In this region a use of the statistical crack growth data is required. Only the primary cracks should be used for determining when a transition from small to large crack growth behavior has occurred. The objective of this section is to show how these crack data can be used to determine when the transition has occurred. The transition from small to large crack growth behavior occurs when crack growth rates for a given loading conditions are the same for equal crack lengths. This occurs when a crack front passes through enough grains to average out the effects of local micro-structural variations. The primary crack growth data were used by Cappelli et al. (2008) to determine the critical crack front length. Cubic polynomial regression analyses were performed on the data for each of the test specimens to obtain crack length versus the number of loading cycles. These equations were then differentiated with respect to N to obtain crack growth rates versus crack length for each test. The results were then used to determine when a transition from small crack behavior to long crack behavior occurred. To achieve this objective, standard deviation values of growth rates were computed to establish the growth rate standard deviation versus crack length trend.

Fractographic analyses of fractured surfaces indicated that the cracks grew with approximately semi-circular crack fronts. It follows that the number of grains, n, intersected by the crack fronts can be computed by use of the equation

$$n = \frac{\pi}{d} a \tag{3.14}$$

where a is the radius of the semi-circular crack surface, and d is the average grain size which was determined by a metallurgical examination.

In the early stages of crack growth there was a large amount of scatter in the crack growth rates. As the crack lengths increased, scatter decreased steadily. The observed behavior was consistent with observations that the crack front must intersect enough grains to provide an adequate sampling of the microstructure. The results obtained indicated that for the given alloy a transition from small crack behavior to long crack behavior occurred when the crack front intersected about 13 grains. At this point the standard deviation value of the growth rate was less than 0.1.

It should be noted that Eq. 3.14 cannot always be used to obtain a relationship between n and a. The required equation depends upon the geometry of the crack.

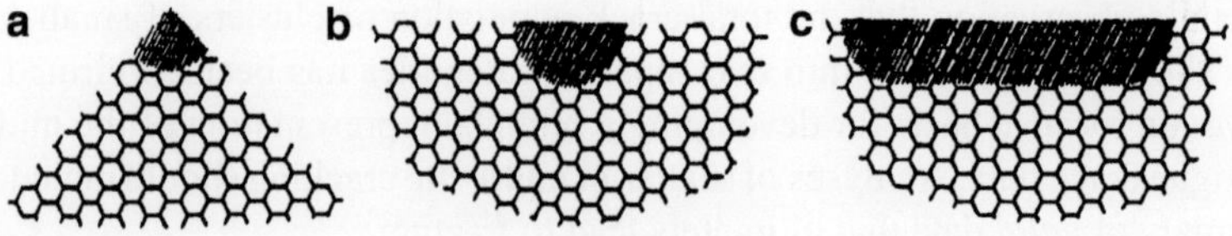

Fig. 3.13 Sample crack geometries with simulated microstructures: (**a**) corner crack. (**b**) A thumbnail crack. (**c**) A surface crack in a plate in bending

Three examples can be used to illustrate the differences. For the test data described here, the semicircular surface crack is appropriate. For a corner crack with the same total surface crack length the number of grains intersected by the crack front would be one half that for which Eq. 3.14 is appropriate. The differences in the number of crack front intersections for these cases are indicated in Fig. 3.13 for three cases in which the microstructures are the same. Note that the crack front of a semi-circular crack intersects twice as many grains as a corner crack of the same depth. The scatter in the growth of a semi-circular can, therefore, be expected to diminish more rapidly than that for a corner crack of the same depth.

When counting grain intersections on a crack front, it is necessary to include only grains that are actively growing. For a crack on the surface of a plate in bending, which is shown in Fig. 3.13c, the relationship is more complex. This type of crack can be created by the initiation and growth of a thumbnail, surface crack into a decreasing stress field. It can, therefore, be expected to grow toward the edges of the plate. Only the intersections of grains on the ends of the crack front should, therefore, be included in a grain intersection count.

3.6 Variable Amplitude Loading

The preceding discussion of fatigue crack growth was concerned with the results obtained from fixed, cyclic loading condition tests. That is, constant amplitude and constant mean load conditions were used to obtain the test results that were presented. Sinusoidal loading conditions were used in those tests. The loading conditions encountered in service can be very complex. Extreme examples of departures from constant amplitude loading occur in aircraft. The type of aircraft and the kind of mission being flown result in characteristic load spectra. These have been monitored to provide to provide load data that can be used in testing programs. Examples of load spectra have been developed for use in the laboratory. These include the following:

- Fighter aircraft lower wing skins- FALSTAFF
- Transport aircraft lower wing skins- TWIST
- Helicopter blades-HELIX, FELIX
- Tactical aircraft cold end engine disks- Cold TURBISTAN

- Tactical aircraft wing skin composites- ENSTAFF
- Tactical aircraft hot end engine disks- Hot TURBISTAN
- Horizontal axis wind turbine blades- WISPER

3.6.1 Tensile Overloads

The transient behavior in fatigue crack growth rate after the application of a single, tensile overload was the subject of experimental studies conducted by Ward-Close and Ritchie (1988) and by McElvily and Yang (1990). Immediately after a single tensile overload was applied, they observed a brief acceleration in the crack growth rate. This was attributed to a crack tip stretching that reduced the closure contact pressure between previously formed obstructions to closure. When the *stretched* crack tip fractured and formed a barrier to closure, there was a transient retardation period during which the crack growth rate decreased.

3.6.2 Compressive Overloads

In the early analyses of overload effects it was a common practice to exclude the effects of compressive overloads (Bucci 1981). It seemed reasonable to believe that no contributions to crack growth were developed by compressive excursions. Zaiken and Ritchie (1985), however, conducted compressive overload experiments on compact tensile specimens of cast I/M 7150 aluminum specimens. They found that crack growth occurred for loading levels below the threshold stress intensity range after the application of a large compressive overload. They attributed this behavior to a flattening of previously formed asperities that reduced the effective stress intensity range. Test results were obtained for three alloys and reported by Carlson et al. (1993). In these tests a period of tensile loading with $R = 0.1$ was interrupted, and compressive loading with $R = -2$ was introduced. After a short interval, the loading ratio of $R = 0.1$ was resumed. The results of the loading excursions for Waspoloy, M 50 NiL and IN-9052 were consistent. They indicated that the crack growth rates were discontinuous at each change in the loading condition, and that the growth rates were substantially larger for $R = -2$ than for $R = 0.1$. When the $R = 0.1$ loading was resumed, the growth rates had the growth rates that were observed at the end of the initial $R = 0.1$ period.

3.6.3 Load Spectra Analyses

Load spectra along with basic fatigue data are integrated into structural design analyses to provide predictions of the fatigue performance of a structure. A comprehensive presentation the details of this process have been presented by Suresh (2002).

It is clear that the loading histories, including the effects of tensile and compressive overloads must be included in these analyses. An early attempt to evaluate some of the proposed analyses was made by designing and implementing a large *round-robin* program Chang and Hudson (1981). Random load spectra for four fighter missions and one transport mission were applied to 2219-T851 aluminum alloy center cracked tension specimens. Different load factors were used to produce a total of 13 test cases. The test data were processed using six different methods of analysis. Two of these included load interaction effects. It was observed that the average of the ratios of predicted to test cycles to failure for 76 analyses was 1.28, with a standard deviation of 0.45. This indicated that a suggestion by Heuler and Schutz (1985) of using a safety factor of 2 was well founded at that time.

3.7 Temperature

The effects of operating temperature are widespread in structural analysis. Examples of this are discussed in Sects. 1.2, 2.3 and 2.4. Extremes of temperature and active atmospheres can also result in a change in the character of the operative mechanisms that affect fatigue crack growth.

Exposure to low temperatures can occur in ships and other vehicles in artic environments and in storage facilities for liquefied gases. Catastrophic failures that have resulted in complete failures have occurred in these environments. The sensitivity of the strength properties of metals in low temperature exposures depends upon the crystal structure. Tchegg and Stanzl (1981) performed fatigue crack tests on both body centered cubic (BCC) and face centered cubic (FCC) metals. They found that the strength and ductility of body-centered cubic alloys, such as ferritic alloys, were quite sensitive to exposures to low temperatures. This was manifested by a behavior in which a reduction in temperature below a 'transition' temperature resulted in a change from ductile to brittle fracture.

The transition temperature can be determined by the use of a Charpy impact test in which a hammer on a pendulum impacts a test specimen with a sharp notch. The distance between the heights of the hammer before and after fracture provides a measurement of the energy absorbed by fracture. Specimens exposed to a range of temperatures are tested to determine the transition from ductile to brittle fracture.

When the ratio of the absolute value of the exposure temperature to the absolute melting temperature, the homologous temperature, of a metal approaches or is greater than 0.5, significant time-dependent, inelastic flow or creep can occur. The mechanism of grain boundary sliding is discussed by Hertzberg (1989), and Raj and Baik (1980) have discussed a mechanism in which cavities are formed in front of a crack tip and crack growth occurs by the successive rupture of the ligaments between cavities.

Cyclic loading at an elevated temperature can, therefore, involve a combination of creep mechanisms and fatigue mechanisms, and coupling effects can be expected to be operative. Although crack growth rate equations are available for creep crack growth and for fatigue crack growth, adding these equations does not account for coupling effects.

3.8 Corrosion

Active atmospheres can introduce various forms of corrosion. Mechanisms that are developed with embrittlement have been discussed by Suresh (2002). These involve: (a) the formation of a hydride phase; (b) enhanced local plasticity; (c) grain boundary weakening; (d) blister and bubble formation. These form of reaction have been observed in zirconium and titanium. The interactions can introduce brittleness, stress corrosion cracking and lead to a reduction in fracture toughness values. Active surface mediums involve gases, aqueous solutions, and liquid metals.

Corrosion-fatigue describes a situation in which corrosive environments promote the initiation and growth of cracks. Small cyclic motion between two metal surfaces can also promote fretting corrosion. This mechanism involves the development of damage in the form of surface micro cracks that can serve as sites for the initiation of fatigue cracks. Examples of fatigue crack growth in corrosive environments, and in fretting corrosion environments have been discussed by Schijve (2001).

3.9 Creep and Fatigue Interaction

The mechanisms for creep and fatigue at elevated temperatures differ. Cyclic loading at elevated temperatures results in a combination of distinct mechanisms. At moderate temperatures, the creep crack growth rate may be of secondary importance. At very high temperatures, creep crack growth may dominate. At intermediate temperatures, both mechanisms can be operative. The proper selection of an alloy may alleviate this potential problem.

At the lower end of the temperature range for creep, and for small stresses, and for high cyclic frequencies, crack growth is cycle dependent. It can be characterized by the use of ΔK.

At high temperatures, high stresses, and low cyclic frequencies, crack growth is time dependent, and creep effects can become dominant. Several parameters have been proposed for characterizing the state adjacent to the crack tip in a body subject to creep. These are based on an evaluation of $C(t)$ which is analogous to the J integral introduced by Rice (1968). Bassini and McClintock (1981), Landes and Begley (1976), and Giseke and Saxena (1989) have presented results for a nonlinear viscous law.

The overlapping, intermediate temperature range involves a combination of the two distinct deformation mechanisms. Crack growth rate equations are available for both fatigue and creep. Simply adding these equations does not, however, include the coupling effects of the creep mechanisms and fatigue mechanisms. Nam (2002) has introduced a damage function model based on the nucleation and growth of grain boundary cavities. He has applied the model to test results obtained on an austenitic stainless steel.

Fretting corrosion can occur when two metal surfaces slide relative to one another. Damage in the form of crack nucleation and crack growth can ensue, and lead to failure. Schijve (2001) has listed the operative mechanisms that are involved in fretting corrosion.

1. Interface pressure.
2. Amplitude of relative motion.
3. Materials.
4. Corrosive environment.
5. Cyclic stress level
6. Surface condition.

References

Bassini JL, McClintock FA (1981) Creep relaxation of stress around a crack tip. Int J Solids Struct 17:479–492

Bucci RJ (1981) Development of a proposed ASTM standard test method for near threshold fatigue crack growth measurement and data analysis. In: Hudak SJ Jr, Bucci RJ (eds) ASTM STP, vol 738. American Society for Testing and Materials, Philadelphia, pp 5–28

Cappelli MD, Carlson RL, Kardomateas GA (2008) The transition between small and long fatigue crack behavior and its relation to microstructure. Int J Fatigue 30:1473–1478

Carlson RL, Kardomateas GA (1996) An introduction to fatigue in metals and composites. Chapman & Hall, London

Carlson RL, Blakely E, Kardomateas GA (1993) Effects of negative r values on fatigue crack growth. In: Bailon JP, Dickson JI (eds) Proceedings of the 5th international conference on fatigue and fatigue thresholds. EMAS, Warley, pp 877–882

Carlson RL, Capelli MD, Kardomateas GA (2007) An investigation of the growth of multi-site fatigue cracks. Int J Fract 10(7):329–332

Chang JB, Hudson CM (1981) Methods and models for predicting fatigue crack growth under random loading. In: ASTM STP, vol 748. American Society for Testing and Materials, Philadelphia, pp 3–39

Collins JA (1981) Failure of materials in mechanical design. Wiley-Interscience, New York

Donaldson DR, Anderson WE (1962) Crack propagation behavior of some aircraft materials. In: Proceedings of crack propagation symposium, vol V II. Cranfield College of Aeronautics, pp 375–441

Forth SC, Johnson WM, Seshadri BR (2006) The effect of the laboratory specimen on fatigue crack growth rate. In: Proceedings of the 16th European conference on fracture, Alexandropoulos

Forth SC, James MA, Johnson WM, Newman Jr JC (2007) Anomalous fatigue crack growth phenomena in high-strength steel. In: Proceedings of the international congress on fracture, Turin

Frost NE (1959) Propagation of fatigue cracks in various sheet materials. J Mech Eng Sci 1:151–170

Frost NE, Marsh KJ, Pook LP (1974) Metal fatigue. Clarendon, Oxford

Gerber H (1874) Bestimmung der zulassigen spannungen in eisen-konstructionen. Zeitschrift des Bayerischen Architeckten and Ingenieur-Vereins 6:101–110

Giseke B, Saxena A (1989) Correlation of creep fatigue crack growth rates using crack tip parameters. In: Salama K, Ravi-Chandar K, Taplin DMR, Rama Rao P (eds) Advances in fracture research, vol 1. Pergamon, Oxford, pp 189–196

Goodman J (1899) Mechanics applied to engineering. Longmans-Green, London

Hald A (1957) Statistical theory with engineering applications. Wiley, New York, pp 51–55

Hertzberg RW (1989) Deformation and fracture of engineering materials. Wiley, New York

Heuler P, Schutz W (1985) Fatigue life prediction in the crack initiation and crack propagation stages. In: Salvetti A, Cavallini G (eds) Durability and damage tolerance in aircraft design. EMAS, Warley, pp 38–69

Johnson LG (1964) The statistical treatment of fatigue experiments. Elsevier, London

Landes JD, Begley JA (1976) A fracture mechanics approach to creep crack growth. In: Mechanics of crack growth, vol 590, ASTM STP. American Society for Testing and Materials, Philadelphia, pp 128–148

Liu HW (1963) Fatigue crack propagation and applied stress range. J Basic Eng 85:116–122

McElvily AJ, Yang Z (1990) Fatigue crack growth retardation mechanisms of single and multiple overloads. In: Kitagwa H, Tanaka T (eds) Proceedings of the 4th international conference on fatigue and fatigue thresholds. MCEP, Birmingham, pp 23–36

Miner MA (1945) Cumulative damage in fatigue. J Appl Mech 67(A):159–164

Nam SW (2002) Assessment of damage and life prediction of austenitic stainless steel under high temperature creep-fatigue interaction. Mat Sci Eng 322:64–72, Elsevier

Palmgren A (1924) Die lebensdauer von kugellagern, 2. Verein Deutscher Ingenieure. 68 :339–347

Paris PC, Erdogan F (1963) Critical analysis of crack propagation laws. Trans ASME J Basic Eng 85:528–534

Paris PC, Gomez MP, Anderson WE (1961) A rational analytic theory of fatigue. Trend Eng 13:9–14

Raj R, Baik S (1980) Creep crack propagation by cavitation near crack tips. Metal Sci 14:385–393

Rice JR (1968) A path independent integral and the approximate analysis of strain concentrations by notches and cracks. J Appl Mech 35:379–386

Schijve J (2001) Fatigue of structures and materials. Kluwer, Dordrecht/London

Suresh S (1991) Fatigue of metals. Cambridge University Press, Cambridge

Swain MH (1992) Monitoring small-crack growth by the replication method. In: Larsen, JM Allison JE (eds), Crack test methods, ASTM STP, vol 1149. American Society for Testing and Materials, Philadelphia, pp 34–56

Tchegg EK, Stanzl S (1981) Fatigue crack propagation and threshold in BCC and FCC metals at 77 and 293K. Acta Metall 29:33–40

Ward-Close CM, Ritchie RO (1988) On the role of crack closure mechanisms in influencing fatigue crack growth following tensile overloads in titanium alloys. In: Newman JC, Elber W (eds) Mechanisms in fatigue crack closure, vol 982, ASTM STP. American Society for Testing and Materials, Philadelphia, pp 93–111

Weibull W (1961) Fatigue testing and analysis of results. Pergamon Press, Oxford

Zaiken E, Ritchie RO (1985) On the role of compressive overloads influencing crack closure and the threshold condition for fatigue crack growth in 7150 aluminum alloy. Eng Fract Mech 22:335–348

Chapter 4
Ceramic Materials

Abstract Because of the inherent brittle response of ceramics to tensile loading, their applications have been limited. Their resistance to oxidation at elevated temperatures has, however, attracted interest in the development of composites that can sustain tensile loading.

4.1 Introduction

Ceramics, which are compounds of metallic and nonmetallic materials such solids as oxides, refractory materials, glass and concrete. Many of these have a crystalline structure, and most have ionic or covalent bonding. Low fracture strength and negligible ductility are characteristics of these solids, and they are commonly described as being brittle.

The uses of ceramics are limited by their inability to tolerate the presence of a crack without failing in a catastrophic mode. Some improvements in strength can be achieved by the introduction compressive surface stresses, and the introduction of crack tip shielding mechanisms. The uses of ceramics are of primary interest in high temperature applications. Alumina, for example, is not subject to oxidation while metals such as tungsten and niobium have very high melting temperatures, but are subject to devastating oxidation in air atmospheres.

Another issue that must be considered concerns how pre-existing flaw distributions affect size and stress distribution effects. Statistical analyses have been used to describe these effects. Weibull (1938) introduced a theory in which pre-existing distributions of flaws were considered. He used extreme value distributions to predict the basic features of brittle fracture. His results indicated that fractures of brittle solids are both size and stress state dependent. Test results obtained by Davidenkov (1947) and by Salmassy et al. (1955) have provided evidence in support of the validity of Weibull's analyses. The complexity of the associated analyses can, however, limit applications of brittle materials.

R.L. Carlson et al., *Mechanics of Failure Mechanisms in Structures*, Solid Mechanics and Its Applications 187, DOI 10.1007/978-94-007-4252-9_4,

4.2 Ceramic Matrix Composites

The development of ceramic matrix composites has been result of a desire to provide a class of materials that can be used in very high temperatures applications. This has, however, required the development of effective reinforcing elements that can provide a resistance to tensile fracture. The ability of ceramics to resist deterioration to exposures to elevated temperatures has created an interest in the development of ceramic matrix composites. The primary objectives of these efforts have been to obtain improvements in their resistance to tensile loading. These efforts have required a selection of the types of reinforcing elements to be used and the development of fabrication processes.

4.2.1 Fabrication

Ceramic composites that have been reinforced by fibers, whiskers or particles are being produced. A variety of fabrication procedures have been developed. These have included hot pressed and sintered mixtures of oxide powders reinforced by either particles or whiskers. Laminated ceramic composites have also been fabricated by tape casting techniques that use whiskers for reinforcement. Ceramic composites have been produced by the use of stacking layers that provide improvements in the mechanical and physical properties. Hot pressing provides densification. The technique used is not limited to flat plates, since the flexible tapes can be formed on mandrels to fabricate complex shapes as discussed by Amateau et al. (1993).

Unidirectional and cross-ply composites of continuous silicon carbide fibers in a Si_3N_4 matrix have been fabricated by the use of a pre-preg tape lay-up procedure. Composite pre-forms have been made by use of hot pressing by Yang et al. (1991).

Ceramic matrix composites with continuous fiber reinforcement and whisker-reinforced ceramic composites have been described by Davidge (1989) and by Warren and Sarin (1989).

4.2.2 Fracture and Fatigue Strengths

The objective in the development of ceramic composites has been to produce components that can operate at temperatures that range from 1,000°C to 1,500°C. The emphasis has, therefore, been on the development of fracture strengths and creep resistances at these temperatures. A measure of the achievement of these goals must ultimately be determined by the results of fracture and fatigue tests. That is, the expected success of the reinforcing elements in introducing mechanisms that retard crack growth must be achieved. The emphasis is on the development of useful strengths and creep resistance at these temperatures. Improvements in these properties over that for un-reinforced ceramics are dependent on the effectiveness of crack tip shielding mechanisms. These include crack tip bridging with debonding, fiber pullout and crack deflection cited by Campbell et al. (1990) and by Rodell et al. (1991). Whisker reinforced ceramics have been discussed by Davidge (1989).

The crack growth behavior of reinforced ceramics is similar to that of unreinforced ceramics. Very small changes in load can result in large changes in crack growth rate. Also, for elevated temperature applications details of the creep-fatigue coupling must be considered in evaluations of test data.

The initial flaw distribution in brittle solids forms the basis for determining the probabilities of failure, and it is tacitly assumed that the initial flaw distribution is not modified by cracks formed during loading. This scenario does not apply for reinforced ceramics. During and prior to final fracture, progressive non-critical micro-cracks can be developed without resulting in failure. This behavior has been discussed by Wagner (1989) and by Lamon (1992).

References

Amateau MF, Conway JC Jr, Bhagat RB (1993) Ceramic composites: design, manufacture and performance. In: Miravete A (ed) Proceedings of the 9th international conference on composite materials, vol. 2. Woodhead, Cambridge, pp 23–31

Campbell GH, Ruhle M, Dalgleish BH (1990) Whisker toughening. J Am Ceramic Soc 73: 521–530

Davidenkov N (1947) The influence of size on the brittle strength of steel. J Appl Mech 14:63–67

Davidge RW (1989) The mechanical properties and fracture of ceramic composites reinforced and with continuous fibres. In: Friedrich K (ed) Application of fracture mechanics to composite materials. Elsevier Science, Amsterdam, pp 547–569

Lamon J (1992) Probabilistic modeling of damage in brittle materials and ceramic matrix composites. In: Blom AF, Beevers CJ (eds) Theoretical concepts and numerical analysis of fatigue. EMAS, Warley, pp 479–504

Rodell J, Fuller ER, Lawn BR (1991) In situ observations of toughening processes in alumina reinforced with silicon carbide whiskers. J Am Ceramic Soc 74:3154–3157

Salmassy OK, Duckworth WH, Schwope AD (1955) Behavior of brittle state materials. Technical report 53–50, **1**, Wright Air Development Center, Wright-Patterson Air Force Base, Dayton

Wagner HD (1989) Statistical concepts in the study of fracture properties of fibers and composites. In: Friedrich K (ed) Application of fracture mechanics to composite materials. Elsevier Science, Amsterdam, pp 39–77

Warren R, Sarin VK (1989) Fracture of whisker reinforced ceramics. In: Friedrich K (ed) Application of fracture mechanics to composite materials. Elsevier Science, Amsterdam, pp 571–614

Weibull W (1938) A statistical theory of the strength of materials. Proc Royal Swedish Acad of Engr Sci, 151, Stockholm

Yang JM, Thayer RB, Chen ST (1991) Creep of fiber-reinforced ceramic matrix composites. In: Tsai SW, Springer GS (eds) Proceedings of the 8th international conference on composite materials. Society for the Advancement of Materials and Process Engineering, Covina, pp 23C1–11

Chapter 5
Polymeric Materials

Abstract This chapter deals with the fatigue failure mechanisms, S-N relations and characterization and quantitative approaches for predicting damage evolution in polymeric laminated composites. Data on the relationships for stiffness, strength, fatigue life, residual strength and the extent of damage are summarized.

5.1 Fatigue Failure

Composite laminates exhibit very complex failure mechanisms under static and fatigue loading because of their polymeric nature and the anisotropic characteristics of their strength and stiffness. Fatigue failure is usually accompanied by extensive damage that multiplies throughout the specimen volume, in contrast to the localized formation of a predominant, single crack, common in isotropic metals or brittle materials. The four basic failure mechanisms in composites are: (1) matrix cracking within a layer (2) delamination, (i.e. interply separation) (3) fiber breakage and (4) fiber-matrix interfacial debonding. Fatigue damage consists of any combination of these mechanisms, with ensuing reduced stiffness and strength. Both the type and degree of damage vary widely, depending on material properties, laminations (including stacking sequence), and loading mode. Unlike metals, where fatigue development can exhibit unique damage features (e.g., fatigue striations), damage development under fatigue loading in composite laminates is similar in appearance to that under static loading, except that damage multiplies as fatigue cycles increase at a given stress level.

5.1.1 Matrix Cracking

In multi-directional laminates under in-plane loading, matrix cracking within a layer usually occurs in succession from the weakest layer to the strongest. For example,

R.L. Carlson et al., *Mechanics of Failure Mechanisms in Structures*, Solid Mechanics and Its Applications 187, DOI 10.1007/978-94-007-4252-9_5,

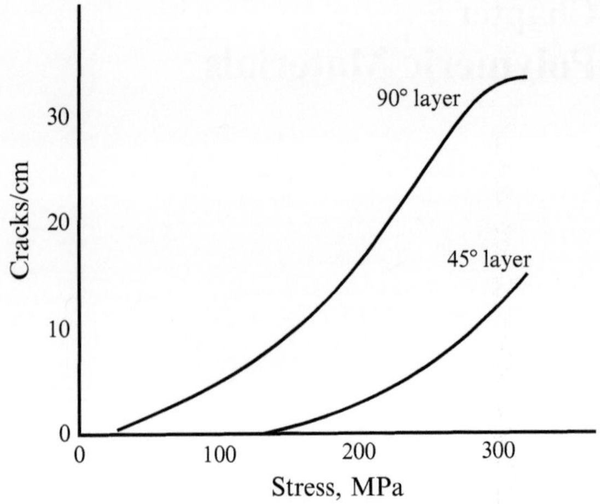

Fig. 5.1 Schematic of the matrix crack density evolution for a [0°/90°/±45°]S laminate under static loading

in a (0°/90°/±45°]S graphite-epoxy laminate subjected to uniaxial tension, the following failure scenario is most likely: The first crack occurs in the 90° layers; with increasing load, more cracks develop, but they are still confined to the 90° layers. As the load increases further, cracks occur in the adjacent 45° layers, appearing at the tips of the 90° cracks in most cases, and extending to the interface of the +45°/−45° layers. Subsequently, the number of cracks increases with the load until final laminate failure takes place. In many cases, laminates reach a crack density limit, after which no new cracks occur before final failure, despite additional loading. The crack density limit for a given layer depends on its thickness and appears to be independent of laminate type. Figure 5.1 shows a schematic of the crack multiplication process for static loading of the [0/90°/±45°]S laminate.

There is a difference in the matrix crack pattern between static and fatigue loading. During fatigue loading, more cracks occur in each layer, and reach a crack density limit, than during static loading. For instance, many cracks in the −45° layers of the [0°/90°/±45°]s laminate occur during fatigue loading, whereas few or none occur during static loading. In addition, many axial cracks initiate at the tips of the transverse cracks and extend along the axial direction as fatigue cycles increase. The multiplication process of transverse cracks during fatigue is shown schematically in Fig. 5.2. The damage ratio is defined as the ratio of the crack density at *n* cycles to the crack density at final failure and is the ordinate axis. Cracks in all off-axis plies are grouped together. Most of the crack multiplication occurs at the first 20% of the fatigue life. A significant amount of fatigue life remains after reaching the crack density limit. With sufficient number of cycles, cracks in the 90° layers of the [0°/90°/±45°]S laminate can be found at almost half the stress at which cracks would appear under static loading.

The 0° layers are also susceptible to cracking in the fiber direction (axial matrix cracks) because of the transverse stress resulting from a mismatch in the Poisson's ratio between different layers. Because this transverse stress is usually small, axial

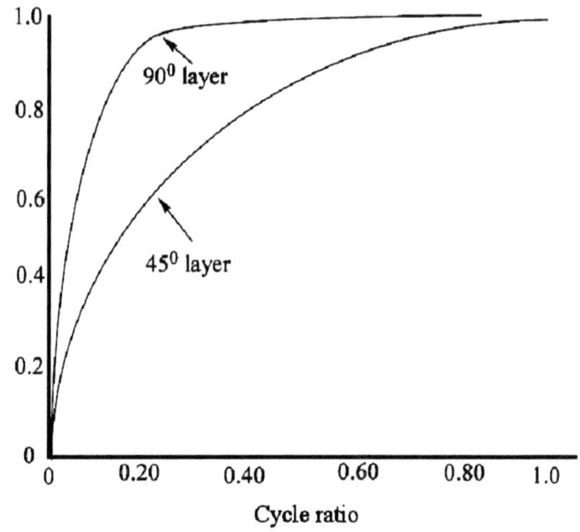

Fig. 5.2 Schematic of the matrix crack density evolution for a $[0°/90°/\pm45°]_S$ laminate under fatigue loading

cracking in the 0° layers may not occur under static loading. However, under fatigue loading, axial cracking can occur, for example, in cross-ply laminates such as $[0°/90°]_S$.

5.1.2 Delamination

There are two types of delamination failure: *free-edge* and *internal.* We shall first discuss free-edge delamination. In 1970, Pipes and Pagano (1970) proved how free-edge interlaminar stress of significant magnitude occurs in composite laminates. With the coordinate through the thickness denoted by z, these interlaminar stress components are the σ_{zz}, τ_{xz} and τ_{yz}. The magnitude and distribution of the interlaminar stress components depend on the composite lay-up, stacking sequence, properties of the constituent materials and the mode of loading (i.e., tension or compression). For example, a cross ply $[0°/90°]_S$ laminate develops less severe interlaminar stresses than a $[\pm30°/90°]_S$ laminate. As a consequence, extensive free-edge delamination is observed in the $[\pm30°/90°]_S$ laminate under applied uniaxial tension, whereas very little or no free-edge delamination has been observed in cross ply laminates. The laminate stacking sequence determines whether the interlaminar normal stress at the free edges is tensile or compressive. As an example, consider the two stacking sequences $[\pm30°/90°]_S$ and $[90°/\pm30°]_S$. The first stacking sequence produces tensile interlaminar stresses at the free edges under applied axial tension, whereas the second stacking sequence produces compressive interlaminar stresses at the same free edges. Thus, the $[\pm30°/90°]_S$ develops extensive free-edge delamination under uniaxial tension, but it does not delaminate under uniaxial compression because of the compressive interlaminar normal stresses produced. For the same reason, free-edge delamination occurs in the $[90°/\pm30°]_S$ laminate under uniaxial compression but not under uniaxial tension.

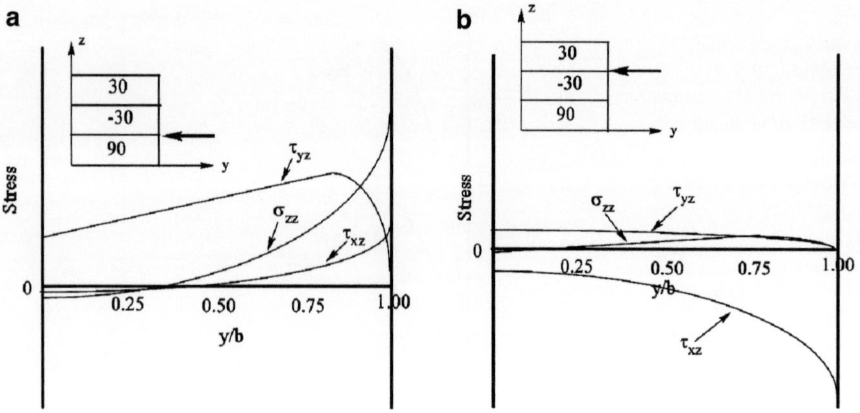

Fig. 5.3 Schematic of the distributions of interlaminar stresses in a $[\pm 30°/90°]_S$ laminates under uniaxial tension (along x). (**a**) Stresses on the 90°/–30° interface (**b**) Stresses on the –30°/+30° interface

The analytical determination of these interlaminar stresses can be done using various techniques (e.g., (Pagano and Soni 1983)). For example, the distributions of σ_{zz}, τ_{xz} and τ_{yz} on the various interfaces of a $[\pm 30°/90°]_S$ laminate under applied uniaxial tension are schematically shown in Fig. 5.3a and b. A large gradient of the interlaminar stress components along the y-direction in the vicinity of the free-edge region exists.

Once free-edge delamination initiates, it typically grows by taking irregular paths at the various ply interfaces; the change in delamination path happens for the most part at the transverse crack tips. In fatigue, the onset of free-edge delamination occurs at stress levels that are smaller than the static ones and typically propagates towards the middle of the specimen as cycles increase. Thus, for every laminate, there exist a fatigue stress level that, within a given number of cycles, the specimen would be delamination-free as long as this stress level is not exceeded.

Although the normal interlaminar stress σ_{zz} is considered the most important in initiating the delamination, the role of the interlaminar stress stresses should not be ignored. As an example, the $[0°/90°/\pm 45°]_S$ laminate, which has a compressive interlaminar normal stress, does not show any delamination under static tension, but under tensile fatigue, considerable delamination occurs at the interface between the +45° and –45° plies. The interlaminar shear stress at the +45°/–45° interface is not large enough to reach the interlaminar shear strength under static loading, but under fatigue loading, the shear stress is enough to initiate a delamination because of the high shear fatigue sensitivity of the epoxy matrix. However, the growth in this case is much slower than the delaminations initiating due to interlaminar tension.

Internal delaminations typically initiate at the ends of transverse cracks (i.e., where the matrix cracks meet the ply interfaces). These are normally confined to short distances, and in general, they are not as significant as the free-edge delaminations, which can initiate more easily and quickly grow through the entire specimen.

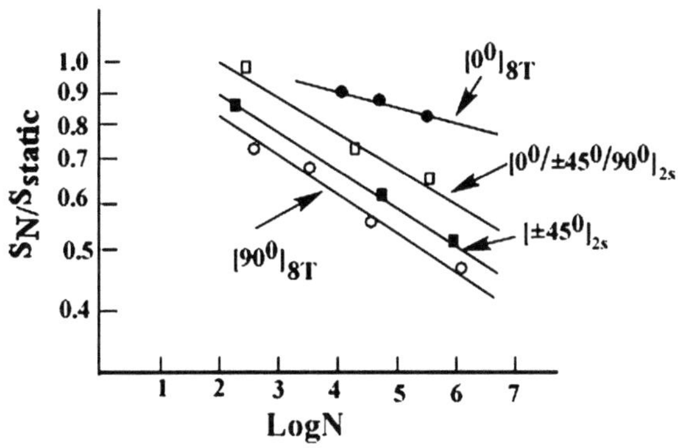

Fig. 5.4 Typical S-N curves for various carbon-epoxy laminates

5.1.3 *Fiber Breakage and Fiber-Matrix Interfacial Debonding*

In most advanced composites, such as boron and graphite fibers with polymeric matrix, the strain-to-failure is greater in the matrix than in the fibers. Therefore, fibers can break (because of defects or localized weakness) before the matrix fails. The crack created by a fiber break grows into the matrix, as load increases, along a path varying with matrix and fiber/matrix interface properties. If fiber/matrix bonding is strong, the crack grows into the matrix, resulting in a fairly smooth fracture surface across the section. With a weak fiber/matrix bond, the crack is more likely to result in interfacial debonding and extensive fiber pull-out. These failure mechanisms can occur under either static or fatigue loading. In the $[0°]_4$ laminate experiments documented in Ryder and Crossman (1983), the 0° fiber breakage was the dominant failure mode both in static and fatigue loading.

5.1.4 *S-N Relations*

In composite laminates, the S-N relation primarily depends on the constituent material properties.

Most advanced fibers are quite insensitive to fatigue, thus the resulting composites show good fatigue resistance. The rule of thumb is that the more off-axis the plies the less the fatigue resistance. Figure 5.4 shows typical S-N curves for a variety of laminates of carbon-epoxy. The ordinate is the ratio of fatigue stress to static strength (note: it is not a linear scale).

The S-N curve for the $[0°]_{8T}$ laminate is relatively flat because the plies have fiber-dominant failure modes. On the contrary, the $[90°]_{8T}$ laminate has matrix-dominant failure modes, thus it is the least resistant to fatigue. Furthermore, the fatigue strength of multi-directional laminates is closely related to the 0° (unidirectional) fatigue strength, with the following empirical relationship

$$\left(\frac{\text{Laminate fatigue strength}}{\text{Unidirectional fatigue strength}}\right) = C\left(\frac{\text{Laminate static strength}}{\text{Unidirectional static strength}}\right)$$

for which the factor C is between 0.85 and 1.0.

5.2 Environmental Effects on Composites

Environmental conditions, such as exposure to water, influence the fibers, matrix material, and the fiber/matrix interface, most often simultaneously. The resulting degradation of the composite may result from (a) loss of strength of the reinforcing fibers by stress-corrosion; (b) chemical degradation of the matrix; (c) induced time and temperature effects on the matrix modulus; (d) loss of adhesion and interfacial bond strength from degradation of the fiber/matrix interface and (e) combined action of temperature and the chemical environment. These factors manifest themselves in an overall reduction of stiffness, which can lead to premature structural instability and/or failure.

Although environmental exposure has an effect on the fibers, the main contributor to the degradation of properties is the matrix. The matrix properties are influenced by the moisture content and temperature variations. Analytical means exist to obtain spatial moisture and temperature profiles in the material and changes in performance due to a specific moisture content and temperature are experimentally determined. Results from a large number of studies on environmental effects on composites have been compiled by Springer (1981).

The important difference between temperature and moisture effects is in the fact that the thermal conductivity is 10^4 to 10^6 times larger than the moisture diffusivity. Thus, the temperature equilibrates much faster than the moisture concentration. A typical variation in moisture content of a four-ply carbon-epoxy composite shows that an equilibrium (saturation) level is attained after many days of exposure.

Figure 5.5 shows the spatial moisture distribution for 2, 10, 20 and 50 days in a 16-ply graphite epoxy which has a moisture exposure of 0.5% on one bounding surface and 1% on the other bounding surface.

Figure 5.6 shows the effect of thickness on the same moisture exposure. Notice that at 100 days the 2 mm thick composite has reached reaches a clear *saturation* state, unlike the 4 and 6 mm cases. Due to the hygroscopic expansion, an important effect of environmental humidity is the generation of curvatures in an otherwise flat panel.

Figure 5.7 shows the curvatures induced simply because of hygroscopic effects (no applied load) for the same configuration as in Figs. 5.5 and 5.6.

More significant is the effect of temperature and moisture on the ultimate tensile strength and stiffness.

An experimental study was conducted by Shen and Springer (1981a), in which the ultimate tensile strengths of Thornel 300/Fiberite 1034 carbon epoxy composites were measured with material temperatures ranging from 200 to 422 K and moisture

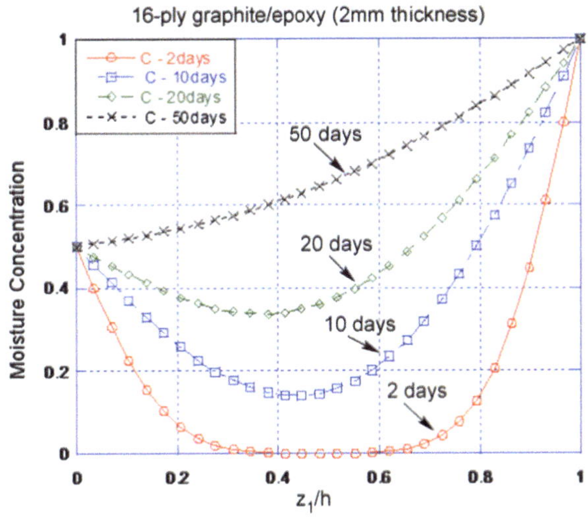

Fig. 5.5 Effect of time: spatial moisture content profile in a 16-ply graphite-epoxy (2 mm thickness)

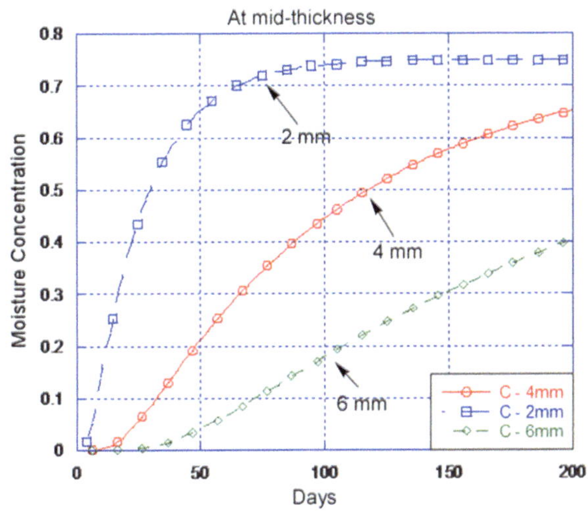

Fig. 5.6 Effect of thickness: moisture diffusion with time for a 2 to 6 mm thick composite

contents from 0% (dry) to 1.5% (fully saturated). The measurements were done on 0°, 45° and 90° laminates and the results are shown in Fig. 5.8a.

The following general conclusions were reached. *Temperature Effects*: (a) For the 0° and 45° laminates, changes in temperature in the range of 200 to 380 K appear to have negligible effects on the ultimate tensile strength, regardless of the moisture content. There might be a slight decrease in strength as the temperature increases from 380 to 450 K. (b) For the 90° laminates, the increase in temperature from 200 to 450 K causes a significant decrease in the ultimate tensile strength.

The decrease depends both upon the temperature and the moisture content and may be as high as 60–90%. *Moisture Effects*: (a) For the 0° and 45° laminates, the

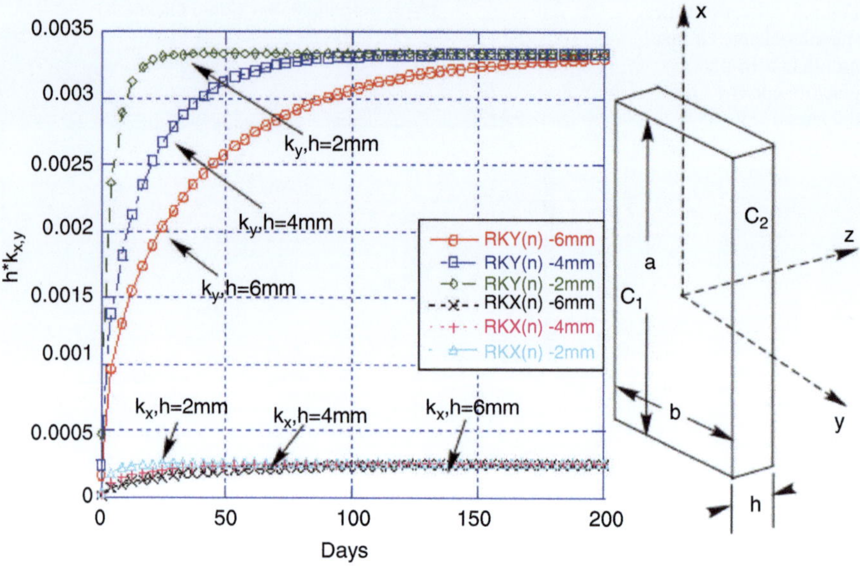

Fig. 5.7 Curvatures induced in a flat panel due to moisture diffusion (no applied load)

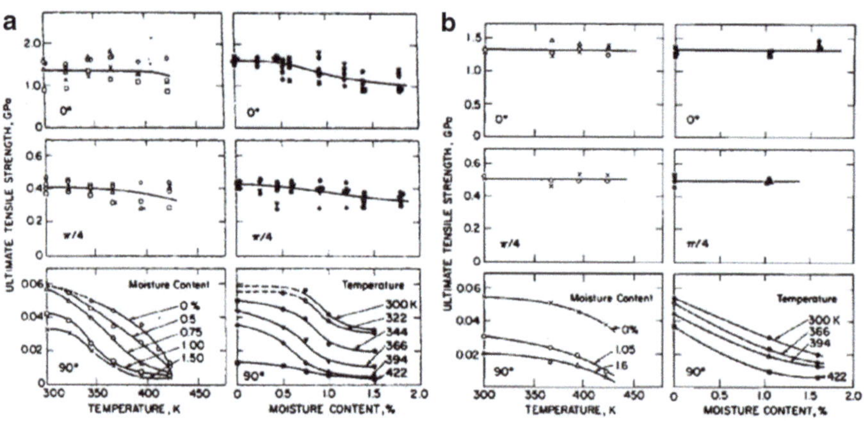

Fig. 5.8 Ultimate tensile strength of carbon epoxy as a function of temperature and moisture content. (**a**) Thornel 300/Fiberite 1034 (Data from Shen and Springer 1981a). (**b**) Hercules AS-5/3501 (Data from Browning et al. 1976)

moisture content has only a small effect on the ultimate tensile strength. Specifically, at moisture contents below 1%, the effects of moisture seem to be negligible. At moisture contents above 1%, the tensile strength decreases with moisture content, the maximum decrease being about 20%. This reduction in strength seems to be insensitive to the temperature. (b) For the 90° laminates, the moisture content affects significantly the ultimate tensile strength and the reduction in strength depends both

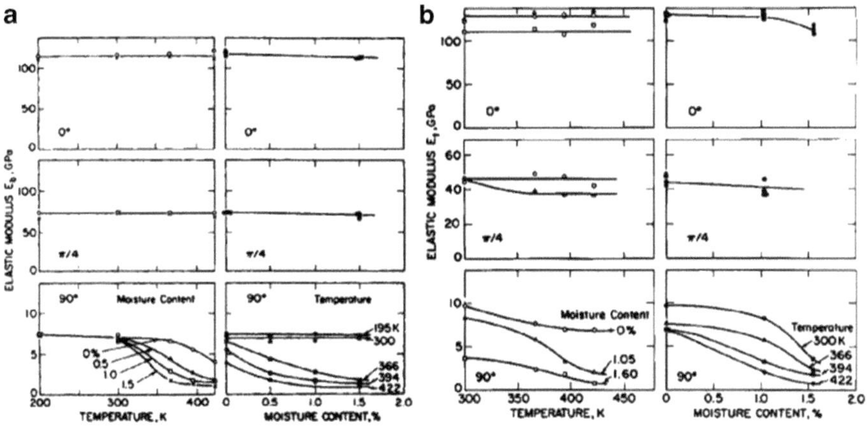

Fig. 5.9 Properties of two carbon epoxies as a function of temperature and moisture content. (**a**) Buckling modulus of Thornel 300/Fiberite 1034 (From Shen and Springer 1981b). (**b**) Tensile modulus of Hercules AS-5/3501 (From Browning et al. 1976)

on the moisture content and the temperature. The reduction in strength may be as high as 60–90%. One other general comment was the observation of occasionally significant scatter in the test data. Similar data were obtained by Browning et al. (1976) on Hercules AS-5/3501 carbon epoxy composite and are shown in Fig. 5.8b.

A companion study was conducted by Shen and Springer (1981b) on the effect of moisture and temperature on the modulus of Thornel 300/Fiberite 1034 carbon epoxy composites and the results are shown in Fig. 5.9a. The following general conclusions were drawn. *Temperature Effects*: (a) For the 0° and 45° laminates, the temperature in the range of 200 to 450 K has a negligible effect on the elastic moduli, regardless of the moisture content. (b) For the 90° laminates, an increase in temperature causes a decrease in the elastic moduli. The decrease in the modulus depends upon both the temperature and the moisture content. For an increase in temperature from 300 to 450 K, the elastic modulus may decrease by as much as 50–90%. *Moisture Effects*: (a) For the 0° and 45° laminates, there appears to be very little change in the elastic moduli over the entire spectrum of moisture content from dry to fully saturated. This conclusion appears to be valid regardless of temperature in the range of 200 to 450 K. (b) For the 90° laminates, the elastic moduli decrease considerably with an increase in the moisture content. The decrease in the modulus depends on both the moisture content and the temperature. The decrease in the value of the modulus may be as high as 50–90%. Again, there was noteworthy scatter in the data of the order of ±20%.

Similar data were obtained by Browning et al. (1976) on the modulus of the Hercules AS-5/3501 carbon epoxy composite and these are shown in Fig. 5.9b. It is apparent that the effects of temperature and moisture on the modulus are similar to those on the ultimate tensile strength (see Fig. 5.8b).

There are other hostile environments for composites and they are dependent on the particular material system. For example, glass fibers are susceptible to attack by acidic environments. This is due to an ion exchange between hydrogen ions in the surrounding medium and alkali ions, in particular sodium, in the glass (Metcalfe and Schmitz 1972). The ion-exchange process is time-dependent, and glass fibers exposed to a low pH environment would eventually fail even at low applied loads, given sufficient time.

References

Browning CE, Husman GE, Whitney JM (1976) Moisture effects in epoxy matrix composites. Composite Materials: Testing and Design. STP-**617** Amer Society for Testing & Materials

Metcalfe AG, Schmitz GK (1972) Mechanism of stress corrosion in e-glass filaments. Glass Tech 12(1):5–16

Pagano NJ, Soni SR (1983) Global-local laminate variational model. Int J Solids Struct 19(3):207

Pipes RB, Pagano NJ (1970) Interlaminar stresses in composite laminates under uniform axial extension. Compos Mater 4:538–548

Ryder JT, Crossman FW (1983) A study of stiffness, residual strength and fatigue life relationships for composite laminates. NASA Report CR-172211, Oct 1983

Shen C-H, Springer GS (1981a) Effects of moisture and temperature on the tensile strength of composite materials. In: Environmental effects on composite materials. Technomic Publishing, Westport, pp 79–93

Shen C-H, Springer GS (1981b) Environmental effects in the elastic moduli of composite materials. In: Environmental effects on composite materials. Technomic Publishing, Westport, pp 94–108

Springer GS (1981) Environmental effects on composite materials. Technomic Publishing, Westport

Chapter 6
Metallic Matrix Composites

Abstract Efforts to make useful metallic composites have required the development of new fabrication procedures. Different approaches have been used and the fabrication techniques adopted are quite distinct from one another. Metal matrix composites have been reinforced by the use of particles, films and fibers.

6.1 Introduction

Dispersion hardening has been long known to provide dramatic increases in strength by the use of distributions of very fine particles that inhibit inelastic slip. Heat treatable aluminum alloys exploit this mechanism. Particle reinforced composites have much higher volume fractions of particles than those used in commercial metals that are classified as dispersion hardened alloys. Also, the size of the particles in metallic composites is much larger. At elevated temperatures, the finer particles dissolve, and no longer provide dispersion hardening. By contrast, aluminum particles have low solubility and the benefits of hardening can be retained at higher temperatures.

Particulate reinforcements in metal matrices have been achieved by the use of molten mixing, powder metallurgy and spray deposition. Components can then be produced by extrusion, rolling or drawing.

Laminated composites can be fabricated by repeated chemical vapor deposition of microscopic films on metallic matrix lamellae. In these composites the lamellar films provide high stiffness and strength and the metal matrix provides ductility and formability (Kreider 1974).

Another example of this class of composites involves eutectic compositions in which two phases solidify in a lamellar sequence. These composites consist of a ductile matrix that is reinforced by a stronger, stiffer lamellar phase described by Thompson and Lemkey (1974).

Considerable effort has also been concentrated on the development of fiber reinforced metallic composites. Emphasis has been directed on the development

R.L. Carlson et al., *Mechanics of Failure Mechanisms in Structures*, Solid Mechanics and Its Applications 187, DOI 10.1007/978-94-007-4252-9_6,

of systems having comparatively low strength matrices reinforced by relatively brittle fibers that have high stiffness and strength. A list of commercially available filaments, a description of manufacturing techniques and their properties has been given by Kreider (1974). Both continuous, aligned fibers and short, chopped fibers have been used to obtain high strength and stiffness. The reinforcement mechanisms of short fibers are analogous to those that are operative for particles and relatively isotropic properties have been achieved.

Continuous, aligned fibers have been used to form laminated sheets that have directional mechanical properties. A variety of techniques have been used to produce these laminates. These include infiltration of molten metal into a fiber network and the use of tapes that can be used to form laminates. Repeated layers can be fabricated to form plates by the use of diffusion bonding. Filament winding procedures can also be used to manufacture axisymmetric bodies. The type of fiber selected for a given metallic composite depends upon the type of service conditions that are to be encountered. The elevated temperature properties of Titanium alloys, for example, can be enhanced by the use of reinforcing filaments as described by Metcalf (1974).

Schijve et al. (1979) have described the fabrication of laminates that consisted of a number of aluminum sheets that were bonded together.

A common requirement for metallic composites is the development of procedures that eliminate voids and porosity since these can serve as origins from which failures can be initiated.

As the technology has evolved, it became clear that the properties of the interfaces determine the mechanical properties. After bond separations, sliding promoted by micro-residual compressive stresses across the interfaces can affect the stress gradients in both the matrix and the fibers.

The compatibility and relative ductility of the particulate must be examined. Micro-residual stress gradients can produce bond fractures or brittle fractures of the particulates. These can serve as sites for the activation of failure mechanisms.

6.2 Experimental and Analytical Results

Schijve et al. (1979) have reported fatigue crack growth results on laminates that consisted of sheets of the 2024-T3 aluminum alloy bonded together. They found that the lives of sheets with a central through-crack were about 60% longer for the laminated sheet that for a solid sheet of the same thickness. They attributed the difference to the fact that the individual thin sheet layers were in states of plane stress for which cracks grow more slowly than in the plane strain states developed in the solid plate.

They also compared the growths of surface cracks for the two cases. The surface cracks in the solid sheet grew more rapidly than in the laminated sheet. In the solid sheet continuous crack penetration and extension occurred. In the composite sheet crack growth extended in the first laminate sheet only, and then formed delaminated branches in the adhesive. The adhesive layers, thus, formed barriers to crack extension.

Schijve (1993) also described the fatigue properties of laminates that consisted of alternate layers of aluminum sheets and glass or of ARAMID fiber combinations. Both unidirectional fiber lay-ups and cross-ply lay-ups have been evaluated in tests. Cross-ply lay-ups are intended for use when tensile and biaxial states are developed.

Experimental and analytical results have indicated that good resistance to crack extension is achieved when partial de-bonding occurs at a fiber interface behind an advancing crack (Hutchinson et al. 1987, 1990; Cox and Marshall 1990; Gupta et al. 1992). If a crack advances through a matrix without de-bonding, the stresses in the fibers will become very large as the crack begins to open. If partial de-bonding occurs, pullout can develop and the stresses in the fibers can be less than the fracture stress. The benefits of the bridging action can then be realized. This introduces tensile crack face forces, and the local stress intensity factor that develops must be subtracted from the global stress intensity factor that is produced by the external loading. The crack face forces depend upon the lengths of the interfacial debonds and the internal friction forces. Ultimately, fibers well behind the crack tip fracture as the crack faces spread apart. The bridging action has been verified by the experimental studies conducted by Sensmeir and Wright (1990), by Barney et al. (1993) and by Davidson (1993) Models describing the effects of bridging on crack tip stress intensity factor in brittle matrix composites have been developed by Budiansky et al. (1986), McCartney (1987) and by Marshall and Cox (1987).

As a crack propagates across a unidirectional metal matrix composite under a tensile load, it most go around fibers. In a multi-layer composite it follows that the crack front will not be straight. In particulate and whisker reinforcements an advancing crack can go over, under, around or through individual particles or whiskers. The advancing crack surface will then be wavy and the crack front will not be straight. These deviations in crack geometry can, therefore, be expected to increase the resistance to crack advance. Rice (1985) and Bower and Ortiz (1990) have performed analytical studies that have examined the consequences of these deviations in crack geometry.

Shang et al. (1988) have studied the roles of these mechanisms in a powder metallurgy aluminum alloy similar to 7091 that was reinforced by a 20% volume fraction addition of fine and coarse silicon carbide particles. They found that the operative mechanisms depended on the particle size and the levels of the cyclic load. At low levels the crack surface roughness was greater for the coarse particulate, and the obstruction to closure reduced the effective range of loading and produced smaller crack growth rates. Although some bridging was observed in the form of matrix ligaments around broken particles, the general conclusion was that fatigue crack growth was not superior to the corresponding alloy.

Huang and Wang (1988) observed that the response of metal matrix composites loaded in compression differed from that for tensile loading. The upper and lower segments of the composite can develop opposite horizontal displacements. This can result in a local kink zone. Analyses designed to exhibit this behavior have been presented by Budiansky and Fleck (1993) and by Slaughter and Fleck (1993).

References

Barney C, Cardova DC, Bowen P (1993) Predictions of crack arrest under cyclic loading in continuous fibre reinforced metal matrix composites. In: Miravete A (ed) Proceedings of the 9th international conference on composite materials. Woodhead Publishing, Cambridge, pp 541–548

Bower F, Ortiz M (1990) Solution of three dimensional problems by the finite perturbation method. J Mech Phys Solids 82:45–57

Budiansky B, Fleck NA (1993) Compressive failure of fiber composites. J Mech Phys Solids 41:183–211

Budiansky B, Hutchinson J, Evans AG (1986) Matrix fracture in fiber-reinforced ceramics. J Mech Phys Solids 34:167–189

Cox BN, Marshall DB (1990) Analogies between bridged cracks in fatigue and fatigue and monotonic loading. In: Kitagawa H, Tanaka T (eds) Proceedings of the 4th international conference on fatigue and fatigue thresholds. MCEP, Birmingham, pp 659–668

Davidson DL (1993) Fatigue crack growth through composites with continuous fiber reinforcement. In: Miravete A (ed) Proceedings of the 9th international conference on composite materials. Woodhead: Cambridge, UK, pp 571–576

Gupta V, Argon AS, Cornie JA (1992) Measurement of interface strength by a laser spallation technique. J Mech Phys Solids 40:141–180

Huang YH, Wang SS (1988) Compressive fatigue damage and associated property degradation of aluminum matrix composites. In: Proceedings of the 4th Japan-U.S. conference on composite materials. American Society for Composites, Technomic Publishing, Lancaster, pp 606–632

Hutchinson JW (1990) Mixed mode fracture mechanics of interfaces. In: Ruhle M, Evans AG, Ashby M, Hirth J (eds) Metal-ceramic interfaces. Pergamon, New York, p 307

Hutchinson JW, Mear ME, Rice JR (1987) An interface between dissimilar materials. Trans Soc Mech Eng 54:828–832

Kreider KG (1974) Introduction to metal matrix composites, vol 4, Composite materials. Academic, London, pp 1–35

Marshall DB, Cox BN (1987) Tensile fracture in brittle matrix composites: influence of fiber strength. Acta Metall 35:2607–2619

McCartney LM (1987) Mechanics of matrix cracking in brittle–matrix fiber–reinforced composites. Proc Royal Soc Lond A409:329–350

Metcalf AG (1974) Fiber-reinforced titanium alloys. In: Kreider KG (ed) Composite materials, vol 4. Academic, London, pp 269–318

Rice JR (1985) First order variations in elastic fields due to variation in location of a planar crack front. J Appl Mech 53:571–579

Schijve J (1993) Development of fiber-metal laminates. New fatigue resistant materials. In: Bailon JPP, Dickson JI (eds), Proceedings of the 5th international conference on fatigue and fatigue thresholds, vol 1. EMAS, Warely, pp 3–20

Schijve J, VanLipzig H, Gestel G (1979) Fatigue properties of adhesively bonded laminated sheet material of aluminum alloys. Eng Fract Mech 12:561–579

Sensmeir MD, Wright PK (1990) The effect of fiber bridging on fatigue crack growth in titanium matrix composites. In: Liawand PK, Gungor MN (eds) Fundamental relationships between microstructure and mechanical properties of metal matrix composites. The Minerals, Metals and Materials Society, Warrendale, pp441–457

Shang JK, Yu W, Ritchie RO (1988) Role of silicon carbide particles in fatigue crack growth in SiC particulate-reinforced aluminum alloy composites. Mater Sci Eng 102:181–192

Slaughter WS, Fleck NA (1993) Compressive fatigue of fiber composites. J Mech Phys Solids 41:1265–1284

Thompson ER, Lemkey FD (1974) Directionally solidified eutectic superalloys. In: Kreider KG (ed) Composite materials, vol 4. Academic, London, pp 101–157

Chapter 7
Biomaterials

Abstract During the twentieth century the use of surgery to correct deformities and defects by the use of implants became quite common. Material implants must satisfy a number of requirements. These include strength, durability and chemical reactivity.

7.1 Introduction

Although body materials such as bone and connective tissue are distinct from such implant materials as metal and plastic, the term biomaterials has been adopted to describe both categories. When implanted, the possibility that chemical reactions between the natural and the implanted materials must be recognized.

Implants include pins, screws and plates for broken bones. Plastic tubes and stainless steel wires are used to repair small bones. Surgeons also replace heart valves and reconstruct knee joints and hip joints. From the above summary it is clear that the technology involved requires specialists in both material science and in the mechanics of failure mechanisms that involve intermittent and varying loads. In view of the consequences of failure of these devices, it is clear that structural integrity is an issue.

7.2 Prosthetic Heart Valves

An average heart beats approximately 30 million times a year. An artificial heart valve that is intended to last for 20 years will, therefore, be exposed to fatigue loading of the order of a billion cycles. The results of an effort to evaluate the fatigue properties of heart valves have been reported by Ritchie and Dauscardt (1990). They conducted experiments designed to determine the fatigue crack growth behavior of the pyrolytic carbon coated graphite material used in prosthetic heart valves.

R.L. Carlson et al., *Mechanics of Failure Mechanisms in Structures*, Solid Mechanics and Its Applications 187, DOI 10.1007/978-94-007-4252-9_7,

Fatigue tests were conducted on compact tension specimens that were made by coating a graphite substrate with a thin layer of silicon carbide.

Crack growth tests under a load shedding procedure were performed to obtain growth rate versus range of stress intensity data. The load ratio used was of 0.1. Tests in both air and in a solution that simulated the environment of a prosthesis implant were conducted. Log-log plots of the crack growth exhibited considerable scatter. Little difference in the results was found between the air and the solution tests.

A threshold ΔK was selected as a value below which crack growth rates did not exceed 10^{-11} m per cycle. This value was 0.7 MPa$\sqrt{m}$. This was approximately 50% of the mean of the values of the critical stress intensity factors that were obtained.

In fitting the data to a Paris type law it was found that the value of the exponent was 18.5. A change of 10% in the range of the stress intensity factor could result in about a sixfold change in the crack growth rate. Since implanted heart valves are not accessible for scheduled inspections, it was concluded that the permissible values of the stress intensity factor range should be well below the threshold value. The authors concluded that the surfaces of an implants should be examined to detect flaws prior to implantation. Also, they recommended that cyclic loading be used.

7.3 Prosthetic Hip Joints

Surgical procedures for using metal prostheses to replace damaged and arthritic joints have been developed to the degree that the operation has become quite commonplace. For the hip joint prosthesis the components are a femur bone, a metal stem and a polymer cement which bonds the stem to the bone. The mechanical properties of these components must be known in order to perform an analysis of the stresses that are developed during various forms of activity.

Morris and Blickenstaff (1976) reported results of early studies on the fatigue fracture of bones. Extended and high levels of activity can cause microscopic damage. Live bone has, however, the capacity to repair or regenerate itself. It has, however, been shown by Carter and Hayes (1976) that if the rate of damage accumulation exceeds the rate of repair, however, fatigue cracks can be developed. Carter and Hayes (1977) also found that fatigue strength decreases with decreasing bone density. This is particularly relevant with regard to prosthesis implants in elderly patients. They also observed that progressive fatigue damage resulted in a decrease in stiffness. This is similar to that observed in fiber reinforced composites. It should be noted that the results reported were on dead bone, so the regenerative function of *live* bone to heal was absent.

An important component in the hip joint prosthesis is the bone cement. The mechanical properties of the cement and its adherence to the bone and to the implant establish the manner in which stress is transferred between the metal stem and the bone. The structural integrity of the cement is crucial to a proper functioning of the remodeled joint.

Rimnac et al. (1986) examined the failure of a metal hip prosthesis and found beach marks and striations typical of fatigue crack growth. Using a finite element analysis they were able to deduce the sequence of loading events that led to failure. Their results led to a change in the prosthesis design and a change from the use of a 316 stainless steel to a cobalt alloy.

One of the primary ingredients in the application of *damage tolerance* concepts is the introduction of scheduled inspections. Since this is difficult in implanted prostheses, improvements must depend upon research and implant experience.

References

Carter DR, Hayes WC (1976) Fatigue life of compact bone-effects of stress amplitude, temperature and density. J Biomechanics 9:27–34

Carter DR, Hayes WC (1977) Compact bone fatigue damages-residual strength and stiffness. J Biomechanics 20:32–37

Morris JM, Blickenstaff LD (1976) Fatigue fractures. Thomas, Springfield

Rimnac CM, Wright TM, Bartel DL (1986) Failure analysis of a total hip femoral component: a fracture mechanics approach. In: Hudson CM, Rich TP (eds) Case histories involving fatigue and fracture mechanics, vol 918, ASTM STP. American Society for Testing and Materials, Philadelphia, pp 377–388

Ritchie RO, Dauscardt RH (1990) Cyclic fatigue crack propagation behavior in pyrolytic carbon coated graphites for prosthetic heart valve applications. In: Kitagawa H, Tanaka T (eds) Proceedings of the 4th international conference on fatigue and fatigue thresholds. MCEP, Birmingham

Chapter 8
Failure in Structural Systems

Abstract Failure in a structural system can arise from a number of causes such as excessive flexibility or unacceptable dynamic response, but an important cause of structural failure is material failure. In previous chapters the failure of different structural materials under a variety of loading conditions is described. The focus in this chapter is on how material failure leads to the failure of a structural system. This, will be examined by the use of three different examples. The first example analyzes failure of a simple planar truss due to either brittle or time independent ductile metal failure, while the second example analyzes failure of a beam due to time independent ductile metal failure. The third example investigates time dependent ductile or creep buckling in an axially constrained column.

8.1 Truss Failure

One of the simplest, practical structures is a planar truss. Trusses are constructed by connecting together a number of prismatic bars or rods at their ends. The end joints are assumed to transmit only axial forces into the bars so that the truss members themselves are subjected solely to axial tensile or compressive loads (with no bending moments). Such joints are difficult if not impossible to achieve in ideal 3D forms, but if the truss members are slender, it can usually be assumed that no matter how the joints are formed, any resulting bending-induced stresses in the bars will be negligibly small compared to the axial stresses.

The most general arrangement of members is a 3D assembly, while a planar truss consists solely of members and applied loads lying within a single plane which is called the plane of the truss. Nearly ideal joints in a planar truss can be achieved using pins or hinges whose axes are perpendicular to the plane of the truss. Such joints are referred to as pinned joints. Even rigidly connected joints can be considered as pinned if the truss members are slender (following the same reasoning noted previously).

R.L. Carlson et al., *Mechanics of Failure Mechanisms in Structures*, Solid Mechanics and Its Applications 187, DOI 10.1007/978-94-007-4252-9_8,

Planar trusses are particularly attractive when they are of sufficiently simple form that the analysis of axial forces in the members and reaction forces at the supports can be carried out using static equilibrium equations alone; such configurations are called *isostatic* or statically determinate. For planar trusses, the isostatic analyses are usually simple enough that they can be carried out efficiently by hand (by analysis of force equilibrium at each joint and/or by analysis of general equilibrium of selected free body diagrams). Moreover, as is the general case for isostatic structures, the calculation of deformation in the truss is entirely independent of the force calculations, regardless of the constitutive behavior of the material used for the members. For this reason, thermal deformation induced by temperature distributions throughout the truss or support deflections affect only the deformation of the truss and do not affect the member forces.

The simplicity of an isostatic, planar truss can be diminished significantly by its failure behavior at extreme levels of loading. Since the truss members are axially loaded bars, the stress state is a simple uniaxial state, and either brittle or ductile metal failure is well defined (failure of a joint or support is more complex and will not be treated in this example). Brittle failure will clearly lead to complete loss of the member. Ductile failure begins with yielding, and if the material has little or no post-yield strength (i.e., an elastic-perfectly plastic material), the member cannot support any increase in load and effectively disappears for further incremental loading. Only if the material has post-yield strength (also called strain hardening) can the member continue to carry axial load.

If any one of the members in the truss should fail to carry its applied axial load, the system will no longer be isostatic and will instead revert to a mechanism and collapse. This is because in an isostatic truss, the number of unknown member forces and support reactions is exactly equal to the number of available equilibrium equations. If one of the truss members can no longer carry load, the number of unknown forces becomes less than the number of governing static equilibrium equations. At this point inertial forces must provide the needed additional unknowns, and so only dynamic equilibrium with consequent motion is possible.

If the material used in the truss members has significant post-yield strength, then the tensile members will continue to carry applied loads until the ultimate stress (and fracture) is reached. However, the reduced member stiffness will lead to increased deformation in the truss and increased joint deflection.

An obvious way to address these shortcomings is to add additional members to the truss so that the loss of a member will not result in collapse of the entire structure. However, this results in a system with more unknown forces than available equilibrium equations and leads to what is called a *hyperstatic* or statically indeterminate structure. Those forces or reactions that exceed the available equilibrium equations are often called the redundant forces or simply redundants, and their number defines the degree of redundancy or hyperstatic degree. Since the equilibrium equations alone are insufficient, it becomes necessary to introduce the analysis of deformation and the material constitutive behavior to solve not only for the member forces and reactions but also for the member extensions and joint deflections. The result is a complete description of the truss under load, but it is achieved at a considerable

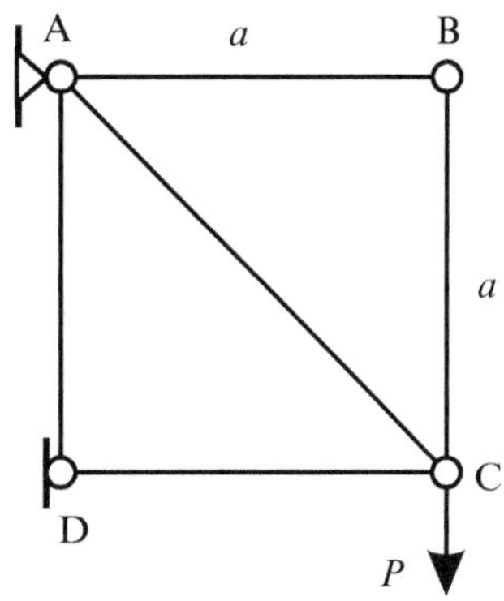

Fig. 8.1 Five-bar isostatic planar truss

increase in the complexity of the required analysis. And while manual calculation methods are practical for structures with only one or two redundants, automated computational methods generally are required for more complex structures.

8.1.1 An Isostatic Truss Example

The effect of these material failure mechanisms on the failure of a truss will be examined in a simple example. Consider a simple isostatic planar truss with five members as shown in Fig. 8.1 above. Assume that the four perimeter members are of length, a, so that the diagonal between joints B and D is $a\sqrt{2}$ in length. Each member has a cross sectional area, A, and elastic modulus, E. The truss is fully supported (pinned to ground) at joint A, while horizontal motion of joint D is constrained. A single downward load, P, is applied at joint C.

From two dimensional force equilibrium at each of the joints (also called the method of joints), the member forces can be computed in terms of the applied load, P, as

$$F_{ABi} = F_{BCi} = F_{ADi} = 0, \quad F_{ACi} = \sqrt{2}P, \quad F_{CDi} = -P \tag{8.1}$$

where the letter i is appended to the subscripts to denote the isostatic truss.

The member extensions can then be computed from their constitutive equations as $e_i = F_i L_i/(EA)_i$ where the subscript identifies the member. For typical structural materials these extensions amount to only about 0.1% of the member lengths, and therefore simple linearized kinematics can be used to compute the joint displacements. More simply, the Unit Load method (Bauchau and Craig 2009) can be used to efficiently calculate selected joint deflections.

As load P is increased, the forces in the four loaded bars will increase proportionally, and these forces result in axial stresses in one or more members that eventually will reach the yield level for a ductile material or the fracture point for

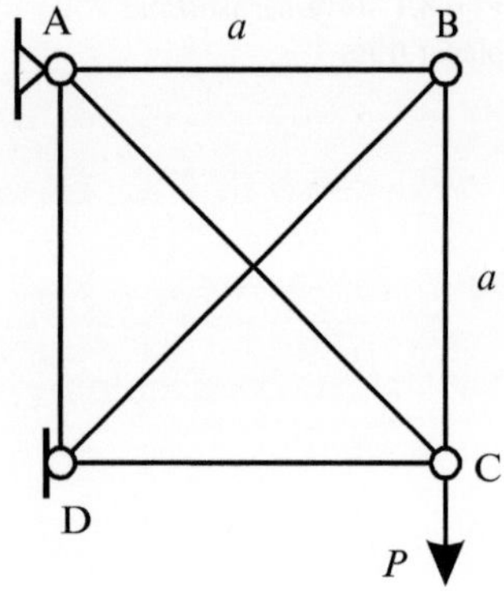

Fig. 8.2 Six-bar hyperstatic planar truss

a brittle material. Member AC clearly carries the greatest load, and it is a tensile load for the downward-acting applied load, *P*. If a brittle fracture occurs, member AC suddenly will no longer carry any load, and without its presence the remaining four members will revert to a simple 4-bar mechanism, leading to collapse of the structure. For ductile behavior, yielding will occur when $F_{ACi} = \sigma_Y A$, or since $F_{ACi} = \sqrt{2}P$ from Eq. 8.1, when the load reaches

$$P_{Yi} = \sigma_Y A/\sqrt{2}. \tag{8.2}$$

Additional increase in the applied load is possible until the post-yield stress in member AC reaches the ultimate stress for the material at which point rupture and loss of the member will occur. If the material is characterized by an elastic-perfectly plastic behavior (zero tangent modulus beyond yield), then member AC can support no further increase in stress beyond the yield stress, and this will lead to arbitrarily large member extension until fracture. In either of these cases, the failure of member AC leads to immediate collapse of the truss as a 4-bar mechanism.

If the truss material exhibits significant post-yield strength, it will continue to carry axial load at stresses above the yield stress. In addition, some indication of pending collapse will be evident in a suddenly increased (nonproportional) joint displacement in the post-yield region. However, in many cases, truss collapse will occur without much if any warning.

8.1.2 A Hyperstatic Truss Example

The isostatic truss examined in the previous example can readily be converted into a hyperstatic configuration by introducing an addition member which introduces an additional unknown internal force. The hyperstatic configuration shown in Fig. 8.2 above is achieved by adding the second diagonal member BD.

There are a number of ways to determine the member forces and support reactions in a hyperstatic system, but all methods must incorporate simultaneous consideration (if not explicit calculation) of the truss deformation. For example, matrix methods or even a simple finite element approach can be used to find the individual joint deflections and then the member forces and support reactions. For simple configurations such as the present case, work and energy principles provide a powerful but very simple approach and will be employed instead. In particular, the Least Work principle (Bauchau and Craig 2009) will be used to directly determine the redundant member force and joint equilibrium will then be used to find the remaining member forces and support reactions (if needed).

The truss has a hyperstatic order of 1 which means there is a single redundant force. It is convenient to choose member AC as the redundant so that $F_{AC} = R$ where R is the unknown (redundant) force in the member. It is then a simple matter of applying the static equilibrium equations (force equilibrium at each joint) to find remaining bar forces as follows:

$$F_{AC} = R, \quad F_{AB} = F_{BC} = F_{AD} = P - \frac{R}{\sqrt{2}}, \quad F_{CD} = \frac{R}{\sqrt{2}}, \quad F_{BD} = -P\sqrt{2} + R. \tag{8.3}$$

Next, the strain energy, U, in the truss due to axial loads in the members is computed by summing the strain energy in each member given by $U_i = \frac{1}{2}{F_i}^2 L_i/(EA)_i$ where the subscript designates the member

$$U = \frac{a}{2EA}\left[\left(3 + 2\sqrt{2}\right)P^2 - 2\left(4 + 3\sqrt{2}\right)PR + 2\left(1 + \sqrt{2}\right)R^2\right].$$

The Least Work principle states that for linear elastic behavior the redundant, R, minimizes the strain energy in the structure. This can be expressed by setting $\partial U/\partial R = 0$ and solving for R to yield $R = \frac{1}{4}(2 + \sqrt{2})P$. Finally, this value of R can be substituted into the equilibrium equations in Eq. 8.3 to determine all the member forces. If necessary, the support reactions can be determined from the additional equilibrium equations at the supports. This results in

$$\begin{aligned} &F_{ABh} = F_{BCh} = F_{ADh} = \frac{1}{4}\left(3 - \sqrt{2}\right)P, \quad F_{CDh} = -\frac{1}{4}\left(1 + \sqrt{2}\right)P, \\ &F_{ACh} = \frac{1}{4}\left(2 + \sqrt{2}\right)P, \quad F_{BDh} = \frac{1}{4}\left(2 - 3\sqrt{2}\right)P \end{aligned} \tag{8.4}$$

where the letter h is appended to the subscripts to denote the hyperstatic truss. These results along with the isostatic truss member forces are tabulated in dimensionless form in Table 8.1. They show that the addition of the additional truss member causes a dramatic redistribution of the forces in the other members. The internal member forces in a truss are generally more uniform in magnitude in hyperstatic compared to isostatic configurations. The support reactions are somewhat less affected.

Table 8.1 Isostatic and hyperstatic truss member forces

Bar	L	$\frac{F_i}{P}$ (isostatic)	$\frac{F_i}{P}$ (hyperstatic)
AB	a	0	$(3-\sqrt{2})/4 = 0.396$
BC	a	0	$(3-\sqrt{2})/4 = 0.396$
CD	a	−1	$-(1+\sqrt{2})/4 = -0.604$
AD	a	0	$(3-\sqrt{2})/4 = 0.396$
AC	$a\sqrt{2}$	$\sqrt{2}$	$(2+\sqrt{2})/4 = 0.854$
BD	$a\sqrt{2}$	–	$(2-3\sqrt{2})/4 = -0.561$

Failure due to yielding will occur in the member with the largest axial load, and in this case it is member AC which is under tension. An even more dangerous situation can also develop in truss members subjected to compressive loading when Euler buckling can occur. This behavior will be discussed in Sect. 8.1.3, and only material yielding will be examined now. From Table 8.1, this will occur when $F_{Ach} = 0.854P = \sigma_Y A$ or when

$$P_{Yh} = \sigma_Y A/0.854 = 1.171\sigma_Y A. \tag{8.5}$$

Compared to the isostatic truss result in Eq. 8.2, this represents a 66% increase in the critical value of the applied load.

In contrast to the isostatic truss, if any of the members in the hyperstatic truss fail, for example due to brittle fracture or to yielding, the remaining members will revert to an isostatic configuration for which member forces can still be determined (from equilibrium equations alone). However, the resulting isostatic truss may then fail due to a sudden increase in the redistributed member forces.

8.1.3 Post-Yield Behavior of the Hyperstatic Truss

As shown in Table 8.1, member AC experiences the greatest load and it is a tensile load. If the truss material is a mild, ductile steel, it is reasonable to assume an elastic-perfectly plastic behavior with no post-yield strength. In this case, once the yield stress is reached in member AC, no further increase in axial stress is possible and the member can undergo an arbitrary extension. The post-yield truss behavior can be analyzed by replacing the tensile force in member AC by a constant force $F_{ACy} = \sigma_Y A$ where A is the member cross sectional area. These two forces are shown in Fig. 8.3 where they act at joints A and C and can be treated just like the applied load, P.

At this point, if the load, P, is increased further, the truss can be analyzed as an isostatic truss with five members and three loads, and the member forces can be found from statics (or by substituting $R = \sigma_Y A$ in Eq. 8.3 for the member forces). It is helpful to plot the dimensionless member forces, $F_i/(\sigma_Y A)$, as a function of the dimensionless load, P/P_{Yi}, and this is shown in Fig. 8.4.

Fig. 8.3 Hyperstatic truss with yielded member AC

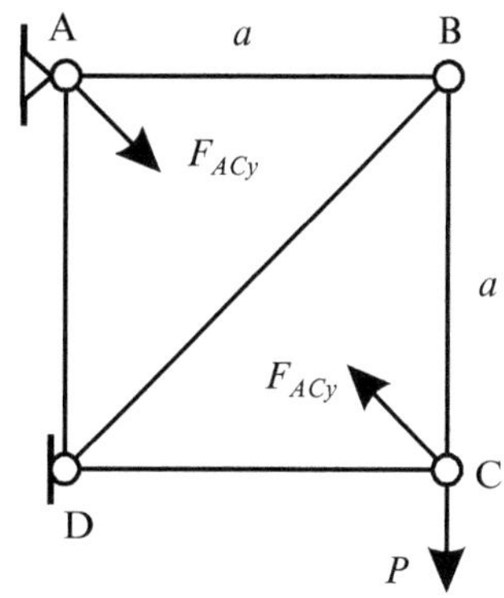

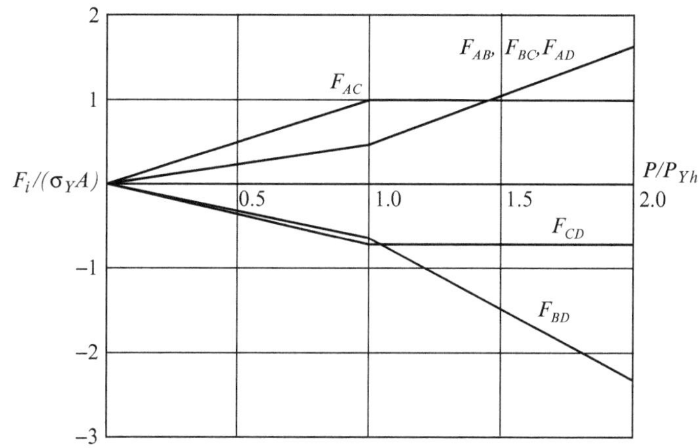

Fig. 8.4 Member forces before and after initial yielding in hyperstatic truss

Yielding in member AC occurs at $P/P_{Yh} = 1$ and limits the load in that member which, in turn, limits the compressive load in CD as well. The redistributed loads in bars AB, BC and AD increase together in tension at a somewhat higher rate while BD increases in compression. When any of the post-yield curves exceeds the ± 1 limits, yielding will occur in that member, and the truss will revert to a mechanism, leading to collapse. This will occur first for bar BD at

$$-1 = \frac{F_{BD}}{\sigma_Y A} = \frac{2 + \sqrt{2} - 4\sqrt{2}\rho}{2 + \sqrt{2}}$$

where $\rho = P/P_{Yh}$. Solving this for ρ yields $\rho = (1+\sqrt{2})/2 = 1.21$ or about 20% above the initial yield load, P_{Yh}. This is the collapse or limit load, and with reference to Eq. 8.2 it can be expressed as

$$P_{Lim} = \frac{1+\sqrt{2}}{2} P_{Yh} = 2P_{Yi}. \tag{8.6}$$

Figure 8.4 also illustrates an important characteristic of hyperstatic structures. Failure of a member in the structure does not cause collapse but instead leads to a redistribution of the member forces through what are referred to as alternate load paths. This will generally avoid a precipitous failure mode, and it results in a more robust structure, albeit a more complex one. It should be noted that in carefully designed hyperstatic structures with higher levels of redundancy (i.e., more members or supports), additional member failures can occur before collapse of the structure.

8.1.4 Buckling of Compression Members in Hyperstatic Truss

An even more dangerous situation can develop in truss members subjected to compressive loading. Under compressive loading, a truss member behaves like a pin-ended (Euler) column and may buckle laterally well before the axial stress approaches yield levels. Such behavior is discussed in Sect. 1.4 in Chap. 1. Assuming pin ended support conditions and a member cross section that is symmetric with respect to the truss plane, the buckling load is the Euler load given by $F_{cr} = -\pi^2 EI/L^2$ where L is the member length and I is the cross sectional area moment of inertia about an axis normal to the plane (the negative sign is used here to denote a compressive load). If all members have the same cross section, then the member with the greatest length will have the lowest buckling load.

For the present example, the diagonals have the greatest lengths, and this leads to a buckling load $F_{cr} = -\pi^2 EI/(2a)^2$. For example, if $a = 1$ m and the members are tubular steel with a square cross section having sides $d = 20$ mm and wall thickness $t = 1$ mm, then $F_{BCcr} = -4,752$N. On the other hand, if a typical yield stress for steel is taken as 400 MPa, then the critical load based on yielding under compression is now $F_{BCcr} = -30.4$kN. Comparing these two results, it is clear that failure due to buckling will occur well before failure due to yielding, and it will likely occur without much warning.

Members BD and CD in the hyperstatic truss are both in compression and are therefore subject to column buckling as described earlier. Since all members are assumed to have the same bending stiffness, *EI*, the ratio of buckling loads in bars CD and BD depends solely on their lengths, and for this example, $F_{CDcr}/F_{BDcr} = (L_{BD}/L_{CD})^2 = 2$. From the hyperstatic analysis summarized in Table 8.1, the ratio of compressive forces in these bars is $F_{CD}/F_{BC} = 1.06$. Since this ratio is

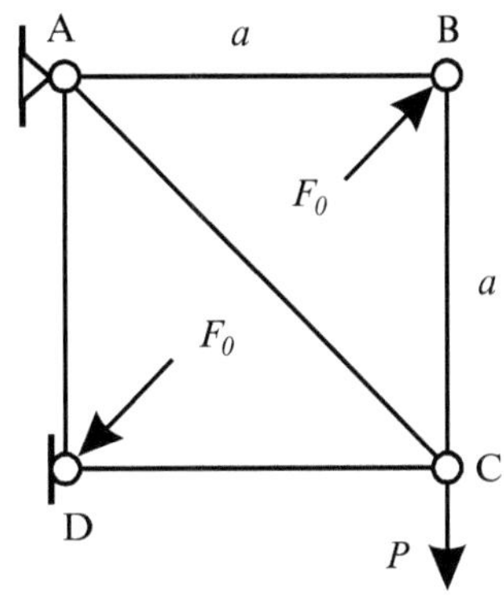

Fig. 8.5 Hyperstatic truss with buckled compression member BD

less than 2, it follows that diagonal bar BD will buckle first, and from the yield versus buckling comparison, this load will typically be considerably less than the yield load.

The initial failure of the hyperstatic truss is due to buckling of bar BD, and by equating the bar force given in Table 8.1 to the buckling (Euler) load, $F_{BDcr} = -\pi^2 EI/L^2$, for member BD with $L = \sqrt{2}a$, the critical applied load is

$$P_{cr} = \frac{2\pi^2}{3\sqrt{2}-2} \frac{EI}{a^2}. \tag{8.7}$$

When bar BD buckles, it experiences an arbitrary lateral deflection, but the axial load remains essentially constant at a value equal to the critical load. In other words, no further increase in the axial load in bar BD is possible. As a result, bar BD can be replaced by two forces, $F_0 = -F_{BDcr}$, one acting at B and the other at D along line BD and pointed towards nodes B and D as shown in Fig. 8.5.

Since the structure with six bars is hyperstatic with degree 1, the buckling of bar BD and its replacement by two loads leaves the structure in an isostatic configuration. If applied load P is increased beyond P_{crit}, the forces in the bars are redistributed and can be computed using statics as follows:

$$F_{AB} = \frac{F_0}{\sqrt{2}}, \quad F_{BC} = \frac{F_0}{\sqrt{2}}, \quad F_{CD} = -P - \frac{F_0}{\sqrt{2}}, \quad F_{AD} = -\frac{F_0}{\sqrt{2}},$$
$$F_{AC} = \sqrt{2}P + F_0, \quad F_{BD} = F_0. \tag{8.8}$$

This result shows that the axial forces in all bars except CD and AC remain constant while bar CD experiences an increasing compressive load and bar AC an increasing tensile load. For a truss with relatively slender members, bar CD will buckle before any yielding occurs, and this will occur at the critical value of applied load given by setting the bar force from Eq. 8.8 equal to the buckling load

$$F_{CD} = -P - \frac{F_0}{\sqrt{2}} = -\pi^2 EI/a^2$$

and solving for load P to yield

$$P_{Lim} = \frac{\left(4+\sqrt{2}\right)\pi^2}{4}\frac{EI}{a^2}. \tag{8.9}$$

At this point bar CD will no longer carry any additional load, reducing the truss to a mechanism and leading to collapse. This defines the limit load for the hyperstatic truss, and by comparing Eqs. 8.7 and 8.9, it is about 52% greater than the load at which the first member (BD) buckles.

These rather substantial improvements in load-carrying ability for the hyperstatic truss are due to the more even distribution of forces in all of the bars in the truss. The ability to carry additional load after buckling of one bar is due to the presence of multiple load paths in the hyperstatic truss, giving it the ability to redistribute loads within the remaining functional members.

8.1.5 Truss Deflections

The deflections of a truss are defined by its joint displacements, and for a planar truss, these are defined by joint displacements within the plane. Calculation of the joint displacements for both isostatic and hyperstatic trusses requires calculation of the member extensions (negative extensions imply shortening) that must then be transformed into joint displacements using the kinematic relations for the truss.

While finite element methods provide a powerful tool for solving this problem, a much easier solution can be achieved for simple planar trusses by using the Unit Load method (Bauchau and Craig 2009). To find the deflection of a joint in a particular direction, all applied loads are first removed from the truss and replaced by a single unit load acting at the joint whose deflection is sought and in the direction of the unknown deflection. The bar forces for this unit load are then determined and are designated by adding a zero to the subscript, i.e., F_{0BD}. The bar forces due to the applied loads themselves are denoted as before, i.e., F_{BD}. The joint deflection is then given by summing the work done in all bars by the unit load forces acting through the real bar extensions

$$d = \sum_{b=1}^{N} F_{0b}\frac{F_b L_b}{(EA)_b}$$

where A_b are the bar cross sectional areas, E_b are their elastic moduli, L_b are their lengths, and index b designates the bar.

When only a single load is applied at a joint, it is particularly simple to compute the deflection in the direction of the load at that joint. In this case the unit load forces in the truss can be obtained by simply setting the applied load equal to unity.

8.1.5.1 Isostatic Truss Example

The load point deflection for the isostatic truss shown in Fig. 8.1 can easily be computed by using the bar forces determined previously and shown in Eq. 8.1, and the unit load forces can be determined by setting $P = 1$ as follows

$$F_{0\,AB} = F_{0\,BC} = F_{0\,AD} = 0, \quad F_{0\,AC} = \sqrt{2}, \quad F_{0\,CD} = -1. \tag{8.10}$$

The load point deflection is then computed using Eq. 8.9 as

$$d = \sum_{b=1}^{N} F_{0\,b} \frac{F_b L_b}{(EA)_b} = \left(3 + 2\sqrt{2}\right) \frac{Pa}{EA}. \tag{8.11}$$

Considering only yielding, the limit deflection occurs at first member yielding when $P = P_{Yi}$ and is given by

$$d_{Lim} = \frac{4 + 3\sqrt{2}}{2} \frac{\sigma_Y a}{E}.$$

For example, if the steel bars have a square tubular cross section of 20 mm width with a wall thickness of 1 mm, length $a = 1$ m, and $\sigma_Y = 400$MPa, this leads to $d_{Lim} = 7.85$ mm.

8.1.5.2 Hyperstatic Truss Example

The load point deflection of the hyperstatic truss in Fig. 8.2 can be determined in the same way by using the previously determined bar forces shown in Table 8.1. The unit load bar forces can be obtained simply by setting P equal to unity to yield

$$\begin{aligned} F_{0\,AB} &= F_{0\,BC} = F_{0\,AD} = \frac{1}{4}\left(3 - \sqrt{2}\right), \quad F_{0\,CD} = -\frac{1 + \sqrt{2}}{4}, \\ F_{0\,AC} &= \frac{1}{4}\left(2 + \sqrt{2}\right), \quad F_{0\,BD} = \frac{1}{4}\left(2 - 3\sqrt{2}\right). \end{aligned} \tag{8.12}$$

Using the Unit Load method, the load point deflection is then given by

$$d = \frac{5 + 3\sqrt{2}}{4} \frac{Pa}{EA}$$

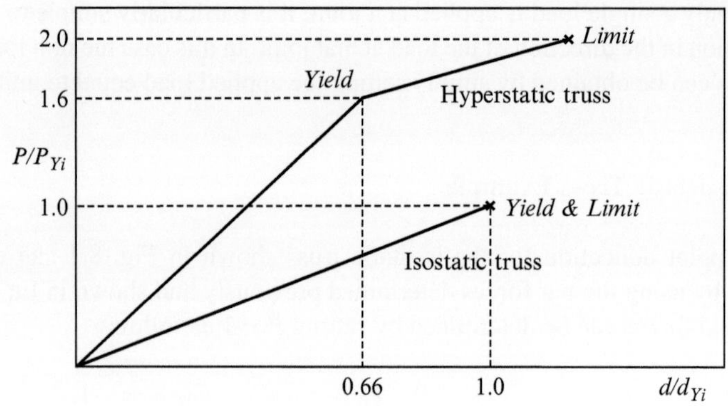

Fig. 8.6 Comparison of isostatic and hyperstatic truss yield behavior

and the deflection when bar AC yields is given when $P = P_{Yh}$ from Eq. 8.5 as

$$d_{Yh} = \frac{5 + 3\sqrt{2}}{2 + \sqrt{2}} \frac{a\sigma_Y}{E}.$$

With the same physical properties used earlier, this results in a deflection, $d_{Yh} = 5.16$mm.

The calculation of the limit deflection when bar CD finally yields is more tedious and requires calculation of the incremental deflection beyond the point when bar AC yields. Since the limit load has already been determined and the post-yield load point stiffness can easily be calculated, it is a straightforward calculation, but it will not be carried out here.

8.1.5.3 Comparison of Yield Behavior

It is interesting to compare the yielding behavior of the isostatic and hyperstatic trusses The result is shown in Fig. 8.6, and the critical points in the figure are determined as follows. The ratios of the critical loads shown on the vertical load axis were determined previously. The ratio of initial slopes can be found from the ratio of initial truss load point stiffnesses, k_h/k_i, given by

$$\frac{k_i}{k_h} = \frac{d_i/P}{d_h/P} = \frac{4}{7}\left(3 + \sqrt{2}\right) = 2.52$$

which indicates that the hyperstatic truss is just over 2.5 times stiffer than the isostatic truss at the load point.

The ratio of the first critical deflection for the hyperstatic truss to the limit deflection for the isostatic truss is given by $d_Y/d_{Lim} = 0.66$. This is plotted on the horizontal deflection axis. The resulting load-deflection curves are piecewise linear as shown.

The addition of the second diagonal member in the hyperstatic truss increases the initial load point stiffness by a factor of just over 2.5 and therefore reduces the load point deflection at the same load. This is achieved by a more uniform distribution of the forces in the members as noted previously.

The hyperstatic truss can continue to carry a load after yielding of member AC which occurs at $P = P_Y$. Beyond this point, the hyperstatic truss reverts to an isostatic with approximately 20% of its original hyperstatic stiffness. The limit deflection occurs at the limit load, P_{Lim}, as shown.

From a practical viewpoint, the hyperstatic truss offers the advantage of being able to accommodate the initial failure by yielding of one of its members without collapse. At this point the apparent truss stiffness is suddenly reduced, but it continues to carry additional load in an isostatic configuration. This provides an obvious precursor to ultimate failure of the structure. On the other hand, the isostatic truss suddenly loses its entire load carrying capability on failure of its first member, and this can occur without any obvious warning.

This comparison of deflections is carried out for the yield behavior. However, as shown previously, buckling of the compression members will occur well before yielding for trusses of practical dimensions. A similar deflection analysis can be carried out for buckling failure, and the load-deflection behavior will be quite similar to that shown in Fig. 8.6.

8.2 Beam Failure

Even though the planar truss discussed in Sect. 8.1 is one of the simplest structures in mathematical terms, the beam is, without question, the most basic practical structure. A beam is defined as a prismatic bar subjected to loads applied transversely. It is likely that the simplest (and maybe the first practical) beam was a slender log or plank laying across a creek and over which a person or animal might walk. Some contemporary examples of beams include the joists used to support floors in buildings, the rafters used to support roofs, as well as the familiar bridges across waterways or roadways. In fact, if a truss member is also subjected to a transverse load, it becomes a beam. It is easy to find many other examples of beams, and they are commonly employed in many structures because their behavior under a wide range of loading is accurately predictable with relatively simple mathematics. Beams under purely axial loads are called columns, but their behavior is quite different and involves a sudden collapse due to excessive lateral deflection called buckling. Analyses and experiments on the buckling of columns and plates are discussed in Chap. 1.

Failure of a beam can involve the development of either excessive lateral deflection or a stress that exceeds the yield stress for the material. Excessive deflection is a matter of serviceability and will not be considered here. Yielding represents a material failure, and the failure can be abrupt if the material is brittle or it can lead to permanent deformation in beams made from ductile materials. In ductile materials, complete failure by fracture will occur when the tensile stress reaches the ultimate value, σ_{ult}, but in highly ductile materials like low carbon steel, this will occur only after considerable inelastic (ductile) deformation. In practice beams are made from ductile materials to avoid brittle failure. The brittle fracture of nominally ductile metals is discussed in Chap. 2.

This section presents a brief summary of beam theory and then examines and compares the behavior of two beams made from a ductile material that are identical except for the manner of support (i.e., the boundary conditions). The first is an isostatic, center-loaded, simply supported beam while the second is a hyperstatic, center-loaded, propped cantilever which is obtained from the first by fully restraining one of its ends.

8.2.1 Beam Theory

Basic beam theory is treated in most introductory structural analysis textbooks such as Bauchau and Craig (2009) or Gere and Goodno (2009). It is based on a fundamental kinematic assumption generally referred to as the Euler-Bernoulli hypothesis which assumes that the most significant deformation of a beam involves the relative rotation of adjacent planar cross sections which rotate but remain planar under deformation. This kind of deformation is caused by the transverse bending moment, M_b, acting at the section, and it causes the beam to bend into a curved shape. If the deflection is contained in a single plane, the problem is defined as two-dimensional or planar beam bending. Purely from symmetry arguments, it follows that this kind of behavior will occur when the plane containing the deflection is also a plane of symmetry for the beam itself. The present treatment will be limited to planar beam bending.

In basic beam theory, the relative rotation of planar cross sections is developed from an assumed linear axial strain distribution over the cross section given by ε. No other strains are considered as significant so it follows that for an elastic material this strain results in an axial stress given by σ (shear stresses may be present on the cross section but they are assumed to result in no significant strains). This is a simple uniaxial stress state. The kinematics leads to a fundamental relationship between the axial strain, ε, and the curvature, κ, given by

$$\varepsilon = -\kappa y \tag{8.13}$$

which is independent of material constitutive relations or equilibrium equations. It should be noted that ε is a linear function of κ at a given y (location) or linear with position y on the cross section at a given κ. For a linearly elastic material with modulus E this leads to a bending stress

$$\sigma = E\varepsilon = -E\kappa y. \tag{8.14}$$

The axial stress distribution must be equivalent to the bending moment applied at the section, and this equivalence can be stated as follows

$$M_b = -\int_{-h/2}^{h/2} y\sigma b\, dy \tag{8.15}$$

where b is the section width and the negative sign is introduced by common sign convention. Using the previous result for σ and carrying out the integration leads to the moment-curvature relation given by

$$M_b = H\kappa \tag{8.16}$$

where the bending stiffness, H, is given by

$$H = \int_{-h/2}^{h/2} Ey^2 b\, dy \tag{8.17}$$

or for a homogeneous material, $H = EI$ where $I = \int_{-h/2}^{h/2} Ey^2 b\ dy$ is the sectional second area moment. Equation 8.16 is the force-deformation relation for a beam and is conceptually equivalent to the familiar $F = ke$ relation for a rectilinear spring where $k =$ stiffness and $e =$ extension. Finally, using κ from Eq. 8.16 in Eq. 8.14 yields the familiar equation for the bending stress in terms of the applied bending moment

$$\sigma = -E\frac{M_b y}{H} = -\frac{M_b y}{I}. \tag{8.18}$$

If it is necessary to determine the beam deflection which is described by the deflection, $v(x)$, in the transverse (y) direction, it is a simple matter to express the curvature as $\kappa = d^2v/dx^2$ which, when substituted into the moment-curvature relation, leads to a simple second order differential equation

$$H\frac{d^2v}{dx^2} = M_b(x)$$

which in turn can be solved for $v(x)$ using appropriate boundary conditions.

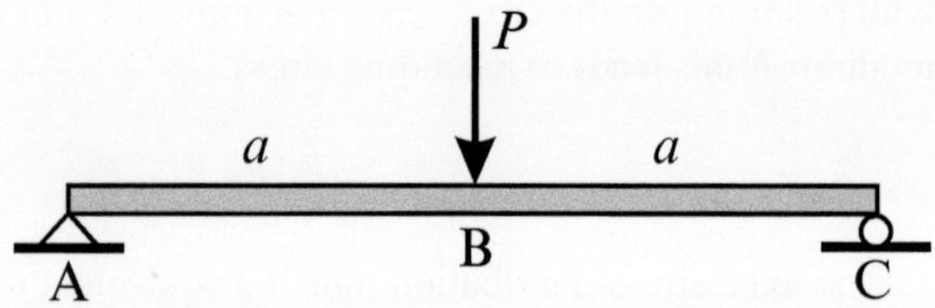

Fig. 8.7 Simply supported beam with midpoint load

8.2.2 *Isostatic Beam Example*

8.2.2.1 Yielding and Post-yield Behavior

Consider a simply supported beam having a rectangular cross section of width, b, and depth, h, and with a midpoint load as shown in Fig. 8.7. The material is assumed to be elastic-perfectly plastic with a yield stress given by σ_Y. For the midpoint loading, the bending moment distribution is triangular and is a maximum at the load point, B, where $M_B = Pa/2$.

Yielding occurs when the bending stress reaches the yield value: $\sigma = \pm\sigma_Y$ (i.e., in either tension or compression). From Eq. 8.18 the point of maximum stress in the beam occurs along the span at the position with the greatest bending moment magnitude, and on that cross section at the location with the extreme values of y, which for the rectangular cross section, means simultaneously at the lower and upper surfaces.

From Eq. 8.18 and using $I = \frac{1}{12}bh^3$ for a rectangular cross section, the bending moment at initial yielding is given by $M_Y = \frac{1}{6}bh^2\sigma_Y$ and the applied load is given by $P_Y = 2\,M_Y/a$. Combining yields

$$P_Y = \frac{bh^2\sigma_Y}{3a}. \tag{8.19}$$

At this point, the stress distribution across the cross section at the midpoint is that shown in Fig. 8.8 below.

As the load is increased further, the axial strain, ε, on the cross section will continue to increase in proportion (because it is due solely to the kinematics and not the material constitutive behavior), but for the elastic-perfectly plastic material, the stress will remain constant at $\pm\sigma_Y$ in those regions where $\varepsilon \geq \pm\varepsilon_Y$ where $\varepsilon_Y = \sigma_Y/E$ is the yield strain. Thus yielded regions where the axial stress $\sigma = \pm\sigma_Y$ will develop at the upper and lower surfaces of the beam (above y_Y and below $-y_Y$) as shown in Fig. 8.8, leaving a region of elastic behavior in between.

As the load is increased even further, the yielded regions will gradually expand towards the centerline, and ultimately they will meet at the centerline so that the axial stress will reach yielding over the entire beam cross section (positive on one

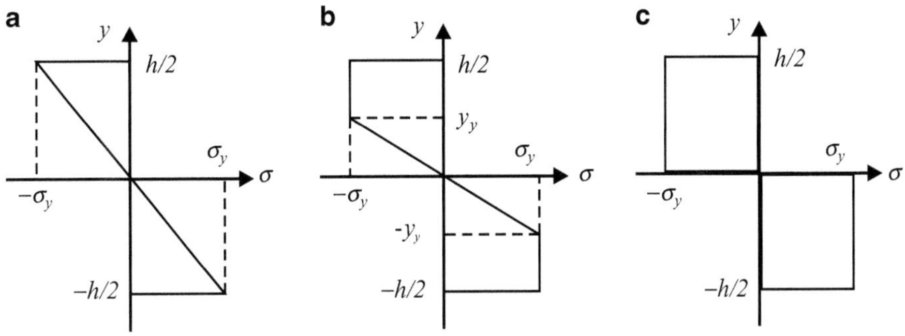

Fig. 8.8 Stress distribution over cross section at beam midpoint. (**a**) Initial yield stress reached on section, (**b**) intermediate condition (**c**) limit state reached on section

half and negative on the other) as shown in Fig. 8.8. At this point, no further increase in stress can occur at this cross section, and the bending moment reaches what is defined as the limiting bending moment, M_{Lim}, at this point.

The axial stress distributions shown in Fig. 8.8 are equivalent to the bending moment acting at the section and can be computed using Eq. 8.15. Applying this to the intermediate stress distribution shown in Fig. 8.8 results in

$$M_b = \frac{1}{4} b h^2 \sigma_Y \left[1 - \frac{1}{3} \left(\frac{y_Y}{h/2} \right)^2 \right]. \tag{8.20}$$

The relationship for the limiting case shown in Fig. 8.8 can be obtained from Eq. 8.20 by setting $y_Y = 0$ and using the result for M_Y to yield $M_{Lim} = \frac{1}{4} b h^2 \sigma_Y = \frac{3}{2} M_Y$ where M_{Lim} is the maximum (or limiting) bending moment that the section can support. At this point, any further attempt to increase the load (and therefore M_b) will be unsuccessful but will result in arbitrarily large relative rotation between adjacent cross sections at point B. In other words, a hinge will develop in the beam at the midpoint under the load. This is generally referred to as a plastic hinge. The load at which this develops can be found from the equilibrium relation between M_B and P given by $M_B = P\,a/2$

$$P_{Lim} = \frac{2 M_{Lim}}{a} = \frac{b h^2 \sigma_Y}{2a} = \frac{3}{2} P_Y \tag{8.21}$$

where Eq. 8.19 is used to introduce the yield load, P_Y.

The development of the hinge can be better understood by constructing a plot of the bending moment (hinge load) as a function of the curvature (hinge rotation) at the midpoint as shown in Fig. 8.9. For moments less than M_Y, the moment-curvature relationship is linear as expected since the beam is entirely elastic. However, once M_b exceeds M_Y, the curvature increases more rapidly and approaches very large values as M_b asymptotically approaches the value $M_b = 3/2\, M_Y = M_{Lim}$.

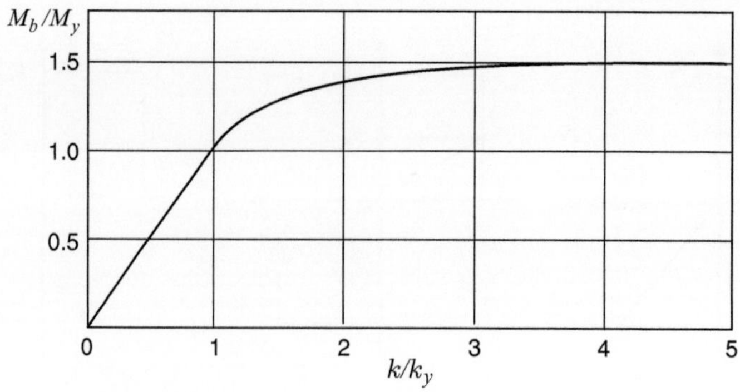

Fig. 8.9 Moment-curvature behavior of plastic hinge in a beam with a rectangular cross section

8.2.2.2 Limit Load Behavior

Returning to the simply supported beam configuration shown in Fig. 8.7, it should be clear that the pinned support at the left end and the roller support at the right end result in an isostatic configuration for which the bending moment can be determined solely from equilibrium considerations. As the limit load, P_{Lim}, corresponding to M_{Lim} is approached, a plastic hinge forms at the midpoint. If the load is increased further so that $P > P_{Lim}$, the plastic hinge will allow arbitrary rotation and, from the kinematics, arbitrarily large deflection, leading to collapse.

It is interesting to compare the behavior of the isostatic beam with the isostatic truss considered previously in Sect. 8.1. For the truss, the yield load is reached when the axial stress in any member reaches yielding, and this leads immediately to collapse of the truss. Therefore for the isostatic truss, the yield load is equal to the limit load and collapse occurs abruptly. On the other hand, once the yield load is reached in the isostatic beam, it will continue to carry 50% more additional load before the limit load is reached. Moreover, the yield condition will be evident by the growing decrease in bending stiffness at the hinge as shown by the reduction in slope in Fig. 8.9. This is due to a redistribution of bending stress within the beam shown in Fig. 8.8 that is not possible in the members of the isostatic truss.

8.2.3 Hyperstatic Beam Example

The simply supported beam in Fig. 8.7 can be changed into a hyperstatic propped cantilever configuration by fixing the right hand end support at C as shown in Fig. 8.10 below. Figure 8.10 also introduces a coordinate system with origin at the left end and identifies the support reactions at A and C. It is necessary to determine the bending moment distribution in the beam in order to begin a failure analysis. However, since there are now three reactions and only two equilibrium equations (vertical force, moment), the configuration is hyperstatic with one redundant quantity which will be taken as M_C.

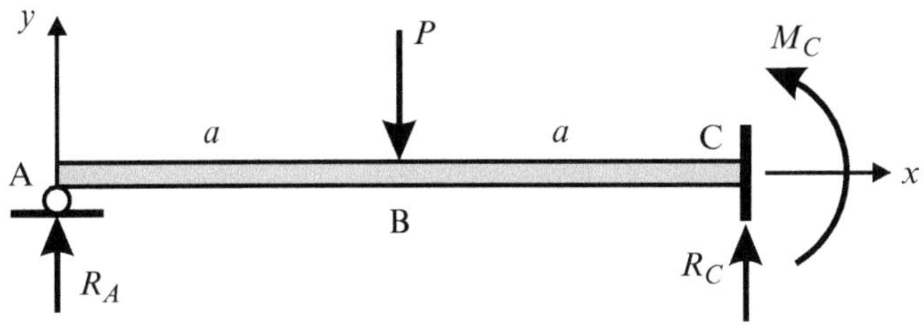

Fig. 8.10 Propped cantilevered beam with midpoint load

The redundant reaction, M_C, can be determined by considering the compatibility and constitutive equations along with the two overall equilibrium equations. There are several well-known methods for doing this, but the Least Work principle (Bauchau and Craig 2009) will be used in this example. The Least Work principle states that for an elastic body, the redundant forces and moments minimize the strain energy which must therefore be expressed in terms of the moment distribution. The two equilibrium equations are $P = R_A + R_C$ and $M_C = 2aR_A - aP$ which can be solved for the two reactions in terms of the redundant moment, M_C , to yield, $R_A = (M_C + aP)/2a$ and $R_C = P - (M_C + aP)/2a$. The bending moment distribution for the two beam segments is then given by $M_{AB}(x) = R_A x$ and $M_{BC}(x) = R_A x - P(x-a)$ and the strain energy is

$$U = \int_0^a \frac{{M_{AB}}^2}{2H}\,dx + \int_0^a \frac{{M_{BC}}^2}{2H}\,dx = \frac{a}{6H}\left(a^2P^2 + 3aPM_C + 4{M_C}^2\right).$$

The Least Work principle requires $\partial U/\partial M_C = 0$ which can be solved for $M_C = -\frac{3}{8}Pa$. As expected, the bending moment distribution shown in Fig. 8.11 assumes extreme values at the supports and at B where the concentrated load is applied.

8.2.3.1 Initial Yielding

The bending moment is the largest at the fixed support C on the right, $M_C = -\frac{3}{8}Pa$, and this is where initial yielding will occur. Since the cross section is the same as before, it follows that the yield moment, M_Y, will be the same as before, $M_C = \frac{1}{6}bh^2\sigma_Y$. The yield load can be found by setting this moment equal to M_C and solving for P_Y to yield

$$P_Y = \frac{4}{9}\frac{bh^2\sigma_Y}{a} = \frac{4}{3}P_{Yi} \tag{8.22}$$

which is 33% greater than the isostatic yield load in Eq. 8.19 simply due to increasing the support fixity at C.

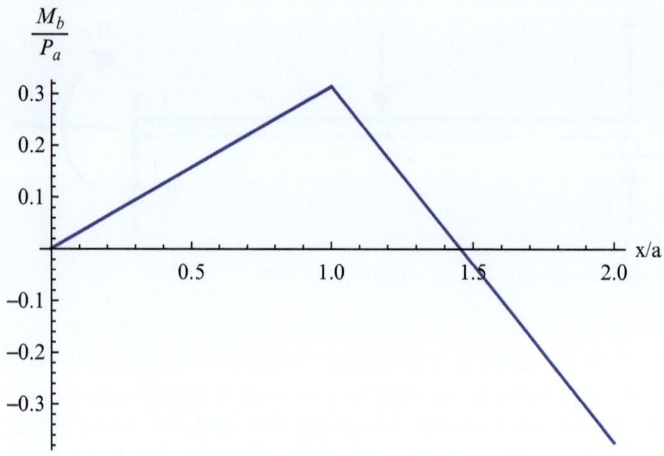

Fig. 8.11 Bending moment distribution on hyperstatic beam

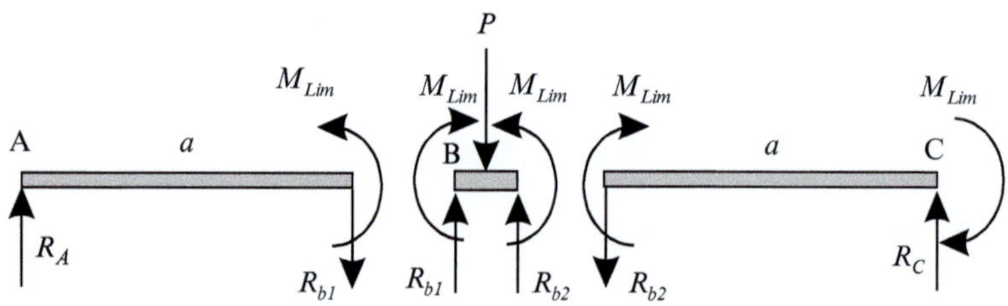

Fig. 8.12 Limit load equilibrium model

8.2.3.2 Limit Load Behavior

In addition to providing a 33% increase in the yield load, the hyperstatic beam will also exhibit the same behavior shown in Fig. 8.9 if the load is increased further. This will lead to a similar limit moment at C at a limit load that is also 33% greater.

However, this is not the true limit load for the hyperstatic system because the resulting plastic hinge at C does not lead to collapse, but instead it reduces the beam to a simply supported configuration. In fact, collapse will not occur until a second plastic hinge is developed, and from Fig. 8.11 this will occur at point B where the bending moment is also a maximum. The relationship between load P and M_{Lim} acting at both plastic hinges can be developed from the equilibrium model shown in Fig. 8.12 above.

There are two equilibrium equations available for each of the three free bodies, and the unknowns are the four reactions shown. End reactions R_A and R_C can be eliminated by writing moment equilibrium at A and C to yield $R_B = M_{Lim}/a$

and $R_B = 2\ M_{Lim}/a$ while vertical equilibrium of the small center segment yields $P_{Lim} = R_B + R_B$. Substituting the first two into the third equation leads to $P_{Lim} = 3M_{Lim}/a$, and using $M_{Lim} = 3/2M_Y$ for a rectangular section leads to

$$P_{Lim} = \frac{3}{4}\frac{bh^2\sigma_Y}{a} = \frac{27}{16}P_Y \tag{8.23}$$

which represents a 69% increase in load beyond the yield load given by Eq. 8.22.

8.2.3.3 Comparison of Limit Loads for Isostatic and Hyperstatic Configurations

The ductile failure of the isostatic beam begins with a well defined yielding at the section subjected to the extreme bending moment. It is followed by collapse due to formation of a plastic hinge after a further 50% increase in load. This behavior can provide a useful safety margin in practical situations.

If additional support is provided to the beam as shown in Fig. 8.10, the result will be a hyperstatic configuration that supports a larger yield load, $P_{Y\,h}$, given by Eq. 8.22, that is 33% greater than the yield load for the isostatic configuration. The post-yield behavior introduces a further safety margin because the initial plastic hinge that will develop at the section subjected to the extreme bending moment does not lead to collapse but instead results in an isostatic configuration. Only when a second plastic hinge develops is the limit condition reached at 69% above the yield load compared to 50% for the isostatic configuration. Finally, it should be noted that the hyperstatic limit load is 50% greater than the isostatic limit load as shown by comparing Eqs. 8.21 and 8.23.

8.2.4 Deflection Analysis

The beam yielding and limit load analyses presented in Sects. 8.2.2 and 8.2.3 can be supplemented by analysis of the beam deflections. Only the isostatic simply supported configuration will be examined, but a similar analysis can be applied to the hyperstatic propped cantilever configuration.

The beam deflection when the yield load is applied can be computed using traditional linearly elastic beam theory. If only the load point deflection is needed, the Unit Load method (Bauchau and Craig 2009) provides a particularly simple formulation in which the point deflection is given by

$$d = \int \hat{M}_b \frac{M_b}{H} dx \tag{8.24}$$

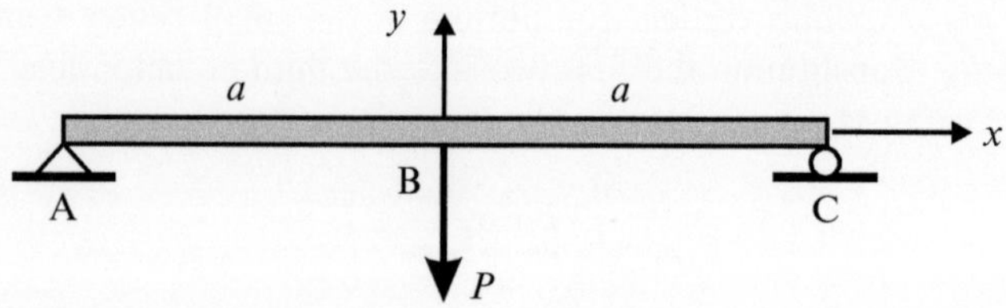

Fig. 8.13 Simply supported beam with center coordinate system

where $\hat{M}_b$ is the bending moment due to a unit load applied at the point where the deflection is to be computed and in the direction of the deflection. The analysis is simplified somewhat by adopting a new coordinate system with origin at the midpoint (B) shown in Fig. 8.13 that takes advantage of the problem symmetry.

8.2.4.1 Deflection at Initial Yielding

The bending moment distribution is symmetric in the coordinate system, and its value for the right half is given by $M_b = M_B(1-\eta)$ where $\eta = x/a$ and $M_B = Pa/2$ is the midpoint moment. The unit load moment distribution is obtained simply by setting $P = 1$, so that the center deflection is given by the integral

$$d = 2\int_0^1 \hat{M}_b \frac{M_b}{H} a d\eta = \frac{1}{6}\frac{Pa^3}{H}$$

where the integral over the right half is doubled to represent the full beam. The deflection at the yield load, P_Y, is then $d_Y = \frac{1}{6}P_Y a^3/H$ and this can be expressed in terms of the yield moment, $M_Y = P_Y a/2$, as

$$d_Y = \frac{1}{3}\frac{M_Y a^2}{H}. \tag{8.25}$$

8.2.4.2 Deflection at Limit Load

Increasing the applied load beyond the yield load leads to inelastic deformation in the region of the beam where the yield stress has been reached. The deflections can be calculated by relatively simple elastic methods provided it can be assumed that no unloading will occur; that is, the analysis will apply to monotonic loading only. In this case, the Unit Load method can be used again, but it will be necessary to recast it in fundamental terms to incorporate the nonlinear "assumed elastic" behavior defined by the elastic-perfectly plastic stress-strain relationship.

This is done by expressing the Unit Load method in its fundamental form which calculates the work done within the beam by the unit load bending moment distribution acting through the real curvature (deformation). Making use of symmetry, this can be written as an integral over only the right half as

$$d = 2\int_0^a \hat{M}_b \kappa \, dx \tag{8.26}$$

where $\kappa(x)$ is the curvature associated with the real deflections. For linearly elastic behavior such as shown in Fig. 8.8, from Eq. 8.16, $\kappa = M_b/H$ which leads to the familiar formula in Eq. 8.24, but for the present analysis, the expression for κ must include the effect of the nonlinear stress distribution shown in Fig. 8.8 for those portions of the beam where the yield stress has been reached. This region begins at point B and extends symmetrically in both directions as the load is increased.

To calculate the midpoint deflection, it is first necessary to determine the yielded region which is the region where $M_b \geq M_Y$. The bending moment is given by $M_b = M_B(1-\eta)$ where $M_B = Pa/2$ is the bending moment at B. At the limit load, $M_B = M_L = 3/2\ M_Y$ so that $M_b = \frac{3}{2}M_Y(1-\eta)$. It follows immediately that $M_b \geq M_Y$ when $\frac{3}{2}M_Y(1-\eta) \geq M_Y$ which is easily solved to yield, $y \geq a/3$. Thus the yielded region is the first third of the half-length on either side of the midpoint (or simply the center third of the complete beam). The Unit Load integral in Eq. 8.26 can be written as

$$d = 2\left(\int_0^{1/3} \hat{M}_b(\eta)\kappa_Y(\eta) a d\eta + \int_{1/3}^{1} \hat{M}_b(\eta)\frac{M_b}{H} a d\eta\right). \tag{8.27}$$

The bending moment due to the real load P is given by $M_b = \frac{1}{2}Pa(1-\eta)$ while the unit load bending moment again is simply the bending moment with the load $P = 1$ which is given by $\hat{M}_b = \frac{1}{2}a(1-\eta)$.

It remains to determine curvature, κ_Y, in the yielded region, and this can be done by consideration of kinematics alone. From the basic beam kinematics of deformation, the strain is always given by Eq. 8.13 regardless of the material behavior. With reference to Fig. 8.8, it follows that the curvature at initial yielding (which occurs in tension at $y = -h/2$) is given by

$$\kappa_Y = \frac{\varepsilon_Y}{h/2} \tag{8.28}$$

where $\varepsilon_Y = \sigma_Y/E$ is the yield strain. However, as the strain in increased as illustrated in Fig. 8.8, it follows that σ_Y is now reached at $y = -y_Y$ so that the curvature can be computed from Eq. 8.13 as

$$\kappa = \frac{\varepsilon_Y}{y_Y}. \tag{8.29}$$

Eliminating the strain ε_Y from Eqs. 8.28 and 8.29 then leads to the desired result

$$\frac{y_Y}{h/2}=\frac{\kappa_Y}{\kappa}$$

and this can be substituted into Eq. 8.20 along with the equation for M_Y to yield

$$M_b=\frac{1}{4}bh^2\sigma_Y\left[1-\frac{1}{3}\left(\frac{y_Y}{h/2}\right)^2\right]=\frac{3}{2}M_Y\left[1-\frac{1}{3}\left(\frac{\kappa_Y}{\kappa}\right)^2\right]$$

which is the desired relationship between the bending moment, M_b, and the curvature, κ. It is necessary, however, to invert this equation which leads to the required result for the curvature within the yielded region of the beam

$$\kappa=\kappa_Y\left(3-2\frac{M_b}{M_Y}\right)^{-\frac{1}{2}}=\frac{M_Y}{H}\left(3-2\frac{M_b}{M_Y}\right)^{-\frac{1}{2}}. \tag{8.30}$$

Substituting Eq. 8.30 along with the expressions for the real and unit load bending moment distributions into the Unit Load integral given in Eq. 8.27 results, after considerable manipulation, in the final result

$$d_{Lim}=2\left[\frac{1}{2\sqrt{3}}\frac{M_Ya^2}{H}\int\limits_0^{1/3}\frac{1-\eta}{\sqrt{\eta}}d\eta+\frac{3}{4}\frac{M_Ya^2}{H}\int\limits_{1/3}^{1}(1-\eta)^2d\eta\right]=\frac{20}{27}\frac{M_Ya^2}{H}. \tag{8.31}$$

With reference to Eq. 8.25, the limit load deflection is over twice the yield load deflection: $d_{Lim}=\frac{20}{9}d_Y=2.22\,d_Y$. Thus while the limit load is 33% greater than the yield load, the peak limit load deflection which occurs at B where the plastic hinge eventually develops is 222% greater. This is due to the asymptotically diminishing rotational stiffness at point B evident in Fig. 8.9 as the plastic hinge is slowly developed. A similar behavior occurs for the hyperstatic configuration. From a practical point of view, this suggests that while design loads should be kept comfortably under the yield load, the design limit load should be kept well below the theoretical results computed in these examples. In this way, elastic behavior will be maintained under normal loading levels and only moderate inelastic deformation will be expected at the design limit load.

8.3 Creep Buckling Failure

The structural system failures analyzed in the previous examples in Sects. 8.1 and 8.2 are not time-dependent (although the failure itself results in dynamic collapse of the structural system). Chapter 1 examines the stability of metallic elements and considers both static and time-dependent material behavior. Section 1.4 in particular introduces creep buckling in which the buckling instability of a column arises because of its time-dependent material properties. This example will examine a simple structural system that includes a column capable of creep buckling but also subjected to an axial constraint. It is based on a paper by Carlson and Morgan (Carlson and Morgan 1981).

8.3.1 Background

The phenomenon of time-dependent buckling due to material creep was recognized first in 1946 in a paper by Ross (1946). Somewhat later Hoff (1958) summarized previous work on this problem in a survey article, and Hult (1966) treated creep buckling as part of a more general treatment of creep in structures. The focus of these studies was on the behavior of a pin-ended column fabricated from materials with different time-dependent constitutive properties. In general, lateral deflection tends to infinity in columns constructed from linear-viscoelastic materials. If either a nonlinear time-dependent or a nonlinear time-independent component of strain is present in the material, then a finite critical time is observed. That is, the lateral deflection increases without bound and the axial load capacity vanishes in a finite time. Huang (1976) pointed out that the small deflection strain-displacement relations used in the previous studies are inconsistent with the fact that large deflections are developed approaching the critical time. In a large deflection analysis, he showed that finite critical times occur only if a lower bound load is exceeded. Thus, while the nonlinear material properties lead to softening effects and produce finite critical times, the incorporation of nonlinear geometrical terms that produce stiffening effects result in longer critical times or even no critical time.

Structural systems such as a truss may consist of many members or a system consisting of a single structural member, such as a beam, may be supported at several locations. The result is that the members are subjected to multiple geometric constraints, and these can significantly modify their behavior as illustrated in the two examples presented in the previous sections. Column buckling is quite sensitive to the presence of lateral constraints which can increase the buckling load by as much as four times. Axial constraint is less effective, but for creep buckling it can have a significant effect on the column behavior as will be shown in an example.

8.3.2 Restrained Column Model

Figure 8.14 shows a pin-ended (simply supported) column with a geometric imperfection defined by an initial lateral deflection $\hat{v}_0(x)$. Axial constraint is provided at the loaded end by a linearly elastic spring of stiffness, k. A compressive axial load is applied by force F, and as the column deforms laterally due to creep, the force

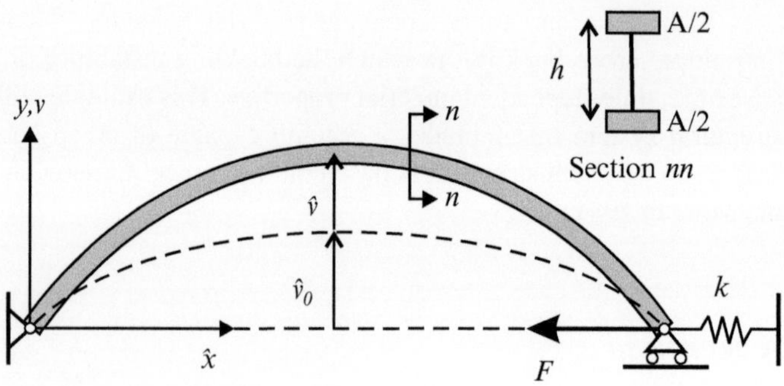

Fig. 8.14 Creep buckling column model

in the spring increases while the force in the column decreases. When $k = 0$, the system reverts to a simple column with no axial constraint. For analysis simplicity, the beam is assumed to have an idealized cross section consisting of two flanges with areas, $A/2$, separated by a web of depth, h, with vanishing thickness.

The constitutive relations for the beam flange material are assumed to be

$$\dot{\varepsilon}_c = \frac{\dot{\sigma}_c}{E} + \left(\frac{\sigma_c}{B}\right)^m$$

$$\dot{\varepsilon}_t = \frac{\dot{\sigma}_t}{E} + \left(\frac{\sigma_t}{B}\right)^m \tag{8.32}$$

where ε and σ are axial strain and stress, respectively, and the dot indicates differentiation with respect to time. The subscripts c and t refer to the compression (lower) flange and the tension (upper) flange. The modulus of elasticity is E while B and m are creep strain parameters.

From consideration of axial force equilibrium at the right hand support, it follows that

$$P = F - kd \tag{8.33}$$

where P is the time-dependent axial compressive force in the column, F is the applied (constant) load, and d is the time-dependent spring extension from the unloaded configuration.

The axial stresses in the two flanges can be determined from equilibrium alone. With reference to Fig. 8.15 which shows a free body from a cut to the left end, axial force equilibrium and moment equilibrium at point Q result in two equations

$$F_c - F_t = P$$

$$F_c + F_t = \frac{2}{h} P\,(\hat{v} + \hat{v}_0)$$

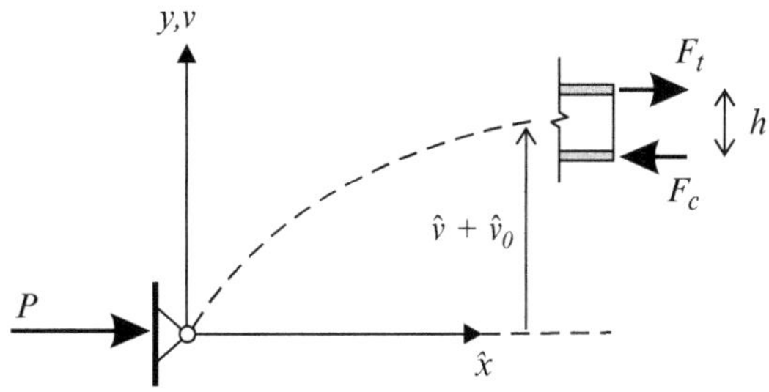

Fig. 8.15 Column flange forces

that can be solved for flange forces F_t and F_c. The flange stresses are then found by dividing by the flange area, *A*/2, and after recasting in terms of the dimensionless lateral deflection, $v = \hat{v}/(h/2)$, results in

$$\sigma_t = -\frac{P}{A}(1 - v - v_0)$$

$$\sigma_c = -\frac{P}{A}(1 + v + v_0). \tag{8.34}$$

If the analysis is confined to only moderately large lateral deflections compared to the beam depth, it is reasonable to use the linear (Euler-Bernoulli) strain-curvature relation, $\varepsilon = -\kappa y$, where $\kappa = d^2\hat{v}/d\hat{x}^2$ is the deflection curvature and $\hat{y} = \pm h/2$ is the flange distance from the beam centerline. For time-dependent behavior, it is necessary to add a time derivative (dot notation) to ε and $\hat{v}$, and after subtracting the expressions for the flange strains and converting to dimensionless deflection, v, and axial position, $x = \hat{x}/L$ this results in

$$\dot{\varepsilon}_t - \dot{\varepsilon}_c = -\frac{1}{2}\left(\frac{h}{L}\right)^2 \frac{d^2\dot{v}}{dx^2}. \tag{8.35}$$

Finally, it is reasonable to assume that the column is inextensionable (i.e., the axial stiffness is much greater than the bending stiffness) so that the end displacement, *d*, under load can be expressed in terms of the deflected shape (Bauchau and Craig 2009) as

$$d = \frac{1}{2}\int_0^L \left[\frac{d(\hat{v} + \hat{v}_0)}{d\hat{x}}\right]^2 d\hat{x} - \frac{1}{2}\int_0^L \left(\frac{d\hat{v}_0}{d\hat{x}}\right)^2 d\hat{x}$$

$$= \frac{h^2}{8L}\int_0^1 \left[\left(\frac{dv}{dx}\right)^2 + 2\frac{dv}{dx}\frac{dv_0}{dx}\right] dx \tag{8.36}$$

where dimensionless deflection, *v*, and axial position, *x*, are used.

At this point, all the necessary equations have been developed, and the next step is to develop the ordinary differential equations for the time-dependent deflection, $v(t)$, and column load, $P(t)$. The expressions for the flange stresses in Eq. 8.34 are substituted into the constitutive law in Eq. 8.32 and the resulting strain rates are substituted into Eq. 8.35 to yield the governing equation for $v(t)$

$$\frac{1}{EA}\left\{-\dot{P}\left[(1-v-v_0)-(1+v+v_0)\right]+2P\dot{v}\right\} \\ -\left(\frac{P}{BA}\right)^m\left[(1-v-v_0)^m-(1+v+v_0)^m\right]=0. \tag{8.37}$$

The column axial load can be found from Eq. 8.33 using Eq. 8.36 for d as

$$P=F-\frac{kh^2}{8L}\int_0^1\left[\left(\frac{dv}{dx}\right)^2+2\frac{dv}{dx}\frac{dv_0}{dx}\right]dx \tag{8.38}$$

and the column load rate, $\dot{P}$, is obtained by taking the time derivative as

$$\dot{P}=-\frac{kh^2}{4L}\int_0^1\left[\left(\frac{dv}{dx}+\frac{dv_0}{dx}\right)\frac{d\dot{v}}{dx}\right]dx. \tag{8.39}$$

Equations 8.38 and 8.39 can be substituted into Eq. 8.37 to yield the final form for the governing equation for $v(t)$ in terms of the fixed load F and the column properties. The result is a rather complicated integro-differential equation, but the basic features of the behavior can be examined using a collocation method described by Hoff (1958).

The column geometric imperfection is assumed to be of the dimensionless form $v_0(x,t) = a_0(t)\sin(\pi x)$, and for a one-term collocation solution, the deflected shape is assumed to be in dimensionless form, $v(x,t) = a(t)\sin(\pi x)$ which satisfies all of the column boundary conditions. Substituting these functions in Eq. 8.37 and collocating at $x = 1/2$ results in the governing nonlinear ordinary differential equation for the deflection amplitude, $a(t)$

$$\frac{1}{EA}\left[2\dot{P}(a+a_0)+2P\dot{a}\right]-\left(\frac{P}{BA}\right)^m\left[(1-a-a_0)^m\right. \\ \left.-(1+a+a_0)^m\right]=\frac{\pi^2}{2}\left(\frac{h}{L}\right)^2\dot{a} \tag{8.40}$$

where from Eqs. 8.38 and 8.39 the column load, P, and load rate, $\dot{P}$, are given by

$$P=F-\frac{\pi^2kh^2}{16L}a(a+2a_0) \tag{8.41}$$

and

$$\dot{P} = -\frac{\pi^2 k h^2}{8L} \dot{a} \left(a + a_0 \right). \tag{8.42}$$

Equation 8.40 with Eqs. 8.41 and 8.42 can be written as a first order ordinary differential equation

$$\dot{a} = f(a) = \frac{G_1(a)}{G_2\,(a))} \tag{8.43}$$

where

$$G_1(a) = \left[\frac{F}{BA} - \frac{\pi^2 k l^2}{16BAL} a \left(a + 2a_0 \right) \right]^m \left[(1 - a - a_0)^m - (1 + a + a_0)^m \right] \tag{8.44}$$

and

$$G_2(a) = -\frac{\pi^2}{8} \frac{kh^2}{EAL} (a + 2a_0)^2 + 2\frac{F}{EA} - \frac{\pi^2}{2} \left(\frac{h}{L} \right)^2. \tag{8.45}$$

8.3.3 Numerical Results

Equation 8.43 can be solved numerically with the initial condition, $\dot{a}(0) = a_o$, using a Runge-Kutta method such as the ODE45 function in Matlab (2011). The geometric parameters used in the solution are $L = 0.3$ m, $h = 5$ mm, $A = 75$ mm^2, and $a_0 = 0.1$. The material properties are taken from Hult (1966) for a carbon steel at 500 C and are $E = 167$ GPa, $B = 100$ GPa-s$^{1/3}$, and $m = 3$. The applied force is assumed to be $F = 8$ kN.

Figure 8.16 shows plots of the lateral deflection dimensionless amplitude, $a(t)$, for two different values of the axial restraint spring stiffness, $k = 0$ (dashed), and $k = 4$ MN/m (solid). Without the spring present, the column buckles at 218 h as evident from the asymptotic increase in lateral deflection. The stabilizing effect of the axial constraint provided by the spring is clear from the behavior of the solid curve.

The form of the lateral deflection curve for $k = 4$ MN/m in Fig. 8.16 suggests the possibility that the deflection rate $\dot{a}(t)$ might go to zero and the column therefore remain in a stable equilibrium under the applied axial load, F. From examination of the governing differential equation, Eq. 8.43, this will occur if $G_1(a) = 0$ or $G_2(a) = \infty$. From Eq. 8.45 it is clear that the second condition cannot happen, however, from Eq. 8.44, the condition, $G_1(a) = 0$ leads to

$$F = \frac{\pi^2}{16} \frac{kh^2}{L} a \left(a + 2a_0 \right)$$

which for a given applied load, F, can be solved for the limiting deflection, $a = a_{lim}$.

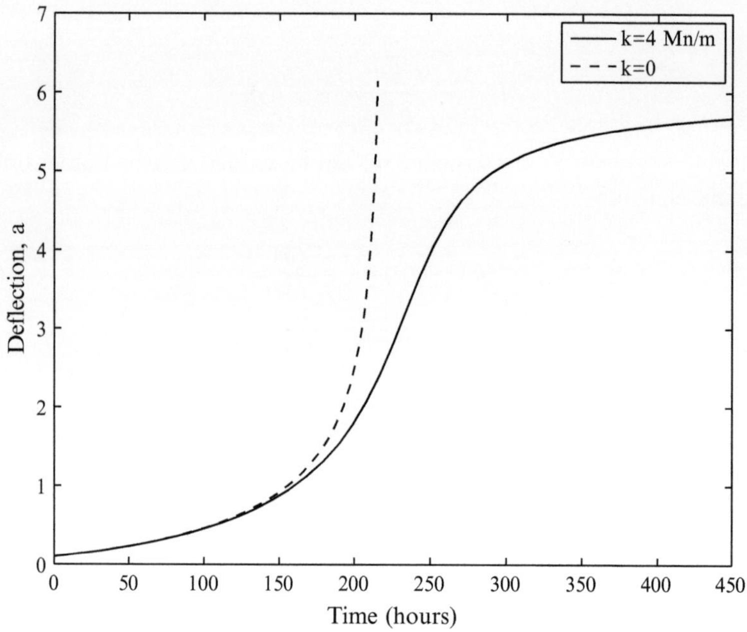

Fig. 8.16 Column deflection time histories with and without axial restraint

The effect of the axial constraint is also evident in the plot of the dimensionless column axial force, *P(t)/F*, shown in Fig. 8.17. This force is readily calculated from the solution for a(t) using Eq. 8.41. The axial force in the unconstrained column remains at the constant value, $P(t)/F = 1$, until collapse at 218 h. On the other hand, with the axial restraint present, the column axial force reduces with time and the force in the spring increases such that equilibrium (Eq. 8.33) is maintained. Thus with the axial spring present, the column axial force decreases continuously and causes a decreasing column deflection rate, which can reach zero and fully stabilize the column.

In this somewhat more complex example, the nonlinear constitutive laws produce a softening effect which leads to a finite collapse time for column creep buckling. However, if nonlinear geometrical effects such as the axial constraint in this example are included, collapse times are either increased or eliminated entirely. As noted earlier, Huang (1976) showed similar effects when nonlinear strain-curvature relations are introduced into the analysis.

Finally, it should be pointed out that while the introduction of axial constraint may stabilize or eliminate creep buckling of a column, other failure modes should be considered. For example, Böstrom (1975) observed that when creep-induced embrittlement processes are operative, fracture and not creep may the mode of failure. Because many heat-resistant alloys are subject to embrittlement, a fracture mode of failure should be considered.

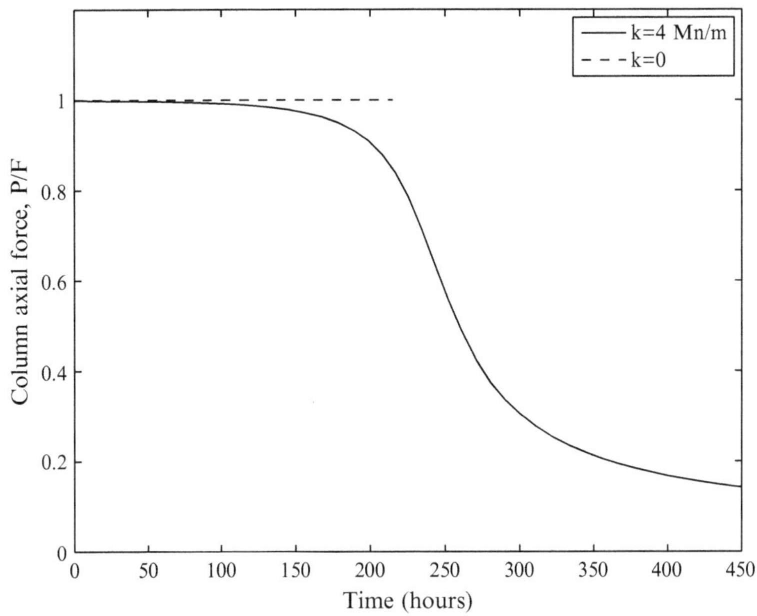

Fig. 8.17 Column axial load time histories with and without axial restraint

8.4 Discussion

The three examples presented in Sects. 8.1–8.3 illustrate how a relatively simple material failure, in this case yielding and eventual fracture of a ductile material such as low carbon steel, or buckling in a slender member, or creep, leads to more complex behavior in structural systems. The simplest ductile behavior considered in these examples is that of an elastic-perfectly plastic material that exhibits a finite elastic stiffness but a zero post-yield stiffness (which, again, is a reasonable characterization for many low-carbon steels).

Planar trusses are the simplest structural systems because they consist of members that transmit only axial forces, and the simplest truss configuration is an isostatic configuration for which equilibrium equations alone define all member forces. For such a truss, loss of any member's ability to carry axial load leads to immediate collapse of the entire structure as a mechanism. However, if additional members are added to create a hyperstatic configuration, it is possible to significantly improve this behavior, although the required analysis is more complex. For such trusses, failure of a single member does not lead immediately to collapse, but instead it reduces the truss to an isostatic configuration that can continue to support the load. It is possible, however, that the redistribution of load that occurs will lead to failure of other members and collapse.

Beam structures with isostatic supports exhibit a progressive failure similar to a hyperstatic planar truss, and collapse does not result from initial yielding, even for the idealized elastic-perfectly plastic material behavior. This is because of the internal redundancy in the beam that allows initial yielding to progressively spread from the extreme edges of a cross section inward until finally the entire cross section has yielded, creating a plastic hinge and leading to collapse. The addition of hyperstatic (redundant) supports eliminates collapse due to a single plastic hinge and leads to a more robust fail-safe configuration.

The third example examines the much more complex situation of column creep buckling which can lead to collapse of the structure. However, the introduction of additional constraint in the form of an axial spring retraining the loaded end of the column can postpone or entirely eliminate the instability and result in a stable, finite deflection.

These examples illustrate that even for relatively simple material failure models such as the elastic-perfectly plastic model, it is possible to design structural systems that are robust in their behavior to this type of failure. A key feature is the ability of the structure to redistribute the internal loads after failure of a member or initial yielding within the structure.

References

Bauchau OA, Craig JI (2009) Structural analysis with applications to aerospace structures. Springer, New York

Böstrom PO (1975) Creep buckling considering material damage. Int J Solids Struct 11(6): 765–775

Carlson RL, Morgan AG (1981) Column creep buckling with end constraint. AIAA J 19(5): 664–666, AIAA 81-4114

Gere JM, Goodno BJ (2009) Mechanics of materials, 7th edn. Cengage Learning, Toronto

Hoff NJ (1958) A survey of the theories of creep buckling. In: Proceedings of the third U.S. national congress of applied mechanics, ASME, New York, pp 29–49

Huang NC (1976) Creep buckling of imperfect columns. J Appl Mech 43:131–136

Hult J (1966) Creep in engineering structures. Blaisdell, Waltham

Matlab (2011) Release R2011a. Mathworks, Natick

Ross AD (1946) The effects of creep on instability and indeterminacy investigated by plastic models. Struct Eng 24(8):413

Index

B
Beam
  bending
    differential equation, 77
    stiffness, 77
    stress, 77
  creep constitutive law, 88
  deflection, 83
    at initial yielding, 84
    at limit load, 84
  Euler-Bernoulli theory, 76
  failure, 75
  hyperstatic initial yielding, 81
  hyperstatic limit load behavior, 82
  hyperstatic yielding, 80
  isostatic limit load behavior, 80
  isostatic post-yield behavior, 78
  isostatic yielding, 78
  plastic hinge, 79
  strain energy, 81
Biomaterials
  prosthetic heart valves, 59
  prosthetic hip joints, 60
Buckling
  column, 4
  column creep, 86
  creep, 5, 6
    in restrained column, 87
  dynamic, 9
  elastic, 6
  Euler load, 5
  Lundquist plot, 8
  Southwell plot, 6
  truss member, 70

C
Ceramics, 41
Composites
  ceramic matrix, 42
  fiber reinforced metallic, 55
  laminated ceramic, 42
  metallic matrix, 55
  particle reinforced, 55
  reinforced ceramic, 41
Compressive overloads, 35
Corrosion, 37
  fatigue, 37
  fretting, 38
Crack(s)
  arrested, 20
  energy absorption, 13
  extension lengths, 16
  growth, 28
  initiation, 20, 30
  long growth regime, 28, 29
  multi-site, 30
  nonpropagating, 20
  primary, 32
  secondary, 32
  small growth regime, 28
  transition, 33
Creep, 86
  and fatigue interaction, 37
Critical energy release rate, 14

D
Damage tolerance, 61
Debonding, 49

R.L. Carlson et al., *Mechanics of Failure Mechanisms in Structures*, Solid Mechanics and Its Applications 187, DOI 10.1007/978-94-007-4252-9,

Delamination
  free-edge, 47
  internal, 48
Design
  damage tolerant, 27
  fail-safe, 26
  safe life, 26
Dispersion hardening, 55

E
Elastic modulus, 1
Embrittlement, 15
Endurance limit. *See* Fatigue limit
Euler load, 5

F
Failure
  beam, 75
  truss, 63
Fatigue, 19
Fatigue limit, 21
Flaws
  initial distribution in ceramics, 43
  pre-existing distribution, 41
Fracture
  brittle, 15
  toughness, 13
    gradients, 15
    testing, 16
Fretting corrosion, 38

G
Goodman
  diagram, 24
  equation, 24

H
Heart valves, 59
Hip joints, 60
Hyperstatic, 64

I
Isostatic, 64

L
Least Work principle, 67, 81
Limit load, 3
Load spectra, 34, 35

M
Modulus
  reduced, 5
  tangent, 5
Modulus of elasticity. *See* Elastic modulus
Multi-site fatigue crack tests, 30

N
Necking, 3

P
Palmgren-Miner rule, 25
Paris region, 29
Plastic hinge, 79
Polymeric composites
  delamination, 47
  environmental effects, 50
  failure mechanisms, 45
  fiber breakage, 49
  fiber matrix interfacial debonding, 49
  matrix cracking, 45
  S-N relations, 49
Polymeric laminated composites, 45
Polymeric materials, 45
Principal structural element, 27
Proportional limit, 2

R
Reversed bending tests, 22

S
Safety factor, 3
S-N curve. *See* Stress-life, diagram
Southwell plot, 6
Strain
  engineering, 2
  true, 2
Stress
  allowable, 3
  engineering, 3
  true, 2
  ultimate, 3
Stress intensity factor, 14, 29
Stress-life
  curves, 21
  diagram, 22
  reversed bending tests, 22
  strategy, 21
Structure
  hyperstatic, 3

integrity, 28
statically determinate (*see* Isostatic)
statically indeterminate (*see* Hyperstatic)
Swain's observation, 33

T
Temperature transition, 36
Tensile overload, 35
Thornel 300/Fiberite 1034, 50
Truss, 63
hyperstatic, 66
deflection, 73
post-yield behavior, 68
isostatic, 65
deflection, 73
yield behavior, 74

U
Unit load method
beams, 83, 85
trusses, 65, 72, 73

W
Williams series, 15

Y
Yield criteria
Tresca, 9
von Mises, 9

MIX
Papier aus verantwortungsvollen Quellen
Paper from responsible sources
FSC® C105338

If you have any concerns about our products,
you can contact us on
ProductSafety@springernature.com

In case Publisher is established outside the EU,
the EU authorized representative is:
Springer Nature Customer Service Center GmbH
Europaplatz 3, 69115 Heidelberg, Germany

Printed by Libri Plureos GmbH
in Hamburg, Germany